Materials *and* Processes *for* Wireless Communications

Related titles published by the American Ceramic Society:

Design for Manufacturability of Ceramic Components (Ceramic Transactions Volume 50)
Edited by Asish Ghosh, Basavaraj Hiremath, and John Halloran
© 1995, ISBN 0-944904-88-2

Manufacture of Ceramic Components (Ceramic Transactions Volume 49)
Edited by Basavaraj Hiremath, Allan Bruce, and Asish Ghosh
© 1995, ISBN 0-944904-87-4

Magnetic Ceramics (Ceramic Transactions Volume 47)
Edited by Bhaskar B. Ghate and John J. Simmins
© 1995, ISBN 0-944904-81-5

Materials Processing and Design: Grain-Boundary-Controlled Properties of Fine Ceramics II (Ceramic Transactions Volume 44)
Edited by Koichi Niihara, Kozo Ishizaki, and Mitsuo Isotani
© 1994, ISBN 0-944904-78-5

Ferroic Materials: Design, Preparation, and Characteristics (Ceramic Transactions Volume 43)
Edited by A.S. Bhalla, K.M. Nair, I.K. Lloyd, H. Yanagida, and D.A. Payne
© 1994, ISBN 0-944904-77-7

Grain Boundaries and Interfacial Phenomena in Electronic Ceramics (Ceramic Transactions, Volume 41)
Edited by Lionel M. Levinson and Shin-ichi Hirano
© 1994, ISBN 0-944904-73-4

Materials in Microelectronic and Optoelectronic Packaging (Ceramic Transactions, Volume 33)
Edited by Hung C. Ling, Koichi Niwa, and Vishwa N. Shukla
© 1993, ISBN 0-944904-63-7

Dielectric Ceramics: Processing, Properties, and Applications (Ceramic Transactions, Volume 32)
Edited by K.M. Nair, J.P. Guha, and A. Okamoto
© 1993, ISBN 0-944904-60-2

Phase Equilibria Diagrams, Phase Diagrams for Ceramists, Volume IX: Semiconductors and Chalcogenides
Edited by Gerald B. Stringfellow
© 1992, ISBN 0-944904-50-5

Materials *and* Processes *for* Wireless Communications

▼▲▼

Edited by
T. Negas
Trans-Tech, Inc.
H. Ling
AT&T Bell Laboratories

▲▼▲

Ceramic Transactions
Volume 53

Published by
The American Ceramic Society
735 Ceramic Place
Westerville, Ohio 43081

Proceedings of the Symposium on Materials and Processes for Wireless Communication, held at the American Ceramic Society Electronics Division Meeting with the International Society for Hybrid Microelectronics, November 15–17, 1994, in Boston, MA.

For information on ordering titles published by the American Ceramic Society, or to request a publications catalog, please call 614-890-4700.

Printed in the United States of America.

1 2 3 4–99 98 97 96 95

ISSN 1042-1122
ISBN 0-944904-94-7

Contents

Preface

Wireless communications technology has exploded over the past 10 years. However, the technology unfolded at a much slower pace as it emerged in the 1960s and 1970s. Its roots lie in now well-established phased array radars allocated to military bands in the GHz range. Later, conventional telephone was linked via land-based microwave systems, and AM radio was coupled to satellites. Evolution through the early 1980s was driven primarily by an infant cellular telephone industry. Today, new consumer applications are conceived, designed, and prototyped at breathless pace, filling numerous commercial bands above ~200 MHz. Major technologies include cellular telephones, pagers, mobile TV transmit, stationary TV transmit/receive, wireless cable, direct broadcasting TV, global positioning, and frequency sensing devices that range from stereo speakers to radar detectors. Interactive TV/computer, automobile collision avoidance, and wireless linkages among phones, computers, cellular systems, buildings, and satellites soon will become reality.

Although many factors have helped spawn and spur industrial growth, two undoubtedly are significant. First, the network analyzer was introduced. This, combined with advancements in computers and software, permitted the analyses and testing of models describing how complex electromagnetic resonant modes couple to and propagate within media. In turn, circuits could be designed, tested, and optimized quickly and accurately. As an important spinoff, new and improved methods needed to characterize the RF and microwave properties of materials became available.

Second, an AT&T group in the 1970s pioneered the first ceramic having the unique combination of high dielectric constant (E_r), very low loss (high Q), and low temperature coefficients of resonant frequency (Tf) and of E_r(TE_r): $Ba_2Ti_9O_{20}$. Once thought improbable, this development opened exciting new avenues. No longer were designs constrained to cumbersome air cavities and transmission lines, lossy capacitor/coil assemblies, and difficult frequency tuning and temperature compensation methods. The ceramic performed all functions. High E_r (~40) reduced circuit/device size and cost. The ceramic could couple to, propagate, and distribute signal with minimal insertion loss while avoiding frequency drift due to temperature variations and to thermal expansion of metal housings.

From the 1980s to the present, new classes of ceramic dielectrics have been developed by industry and by small groups scattered throughout the world. Materials have a broad range of E_r (10–90), Q, and precisely adjustable Tf, thus providing flexibility to designers in terms of need and cost. Most ceramics are used as resonant filters, as metallized coaxial filters or inductors, and as substrates for thick and thin metal film circuits. Since many wireless technologies require circulators and isolators, markets have grown for established magnetic ceramics such as low-loss garnets and ferrite spinels. Future needs of the industry include dielectrics of higher Er and Q, particularly for applications below about 3 GHz. Garnets and ferrites, perhaps with self-biasing, need to be reduced in size and made compatible with silicon and/or GaAs circuitry for applications near and well above 1 GHz.

This volume, the first of its kind, introduces wireless communications to the ceramist. It is a collection of thoroughly reviewed papers selected from the Symposium on Materials and Processes for Wireless Communication, held at the American Ceramic Society Electronics Division Meeting with the International Society for Hybrid Microelectronics (ISHM), November 15–17, 1994, in Boston, MA. The papers are organized into two broad categories: applications and measurements, and designs and materials. (The reader also is encouraged to refer to unpublished presentations from the Symposium on Materials for Wireless Communications, American Ceramic Society Annual Meeting, Cincinnati, OH, 1993.) We thank all speakers, session chairs, and manuscript reviewers for contributing to a refreshing meeting.

T. Negas
H. Ling

Applications and Measurements

The Role of Wireless Communications in The National Information Infrastructure

C. A. Backof, Motorola Land Mobile Products Sector
8000 W. Sunrise Blvd., Plantation FL 33322

ABSTRACT

We live today in a communications world of private voice and public images. The telephone provides private voice contact, while television delivers public broadcast images. With the National Information Infrastructure (NII), that communications model will expand to include private and public access to all forms of information, supporting services and products that can only be imagined. The model is however constrained to fit a wired format, even though people have come to expect wireless access. This paper explores the applications and technologies needed to extend the NII with broadband wireless access, and presents estimates of the spectrum required to achieve this extension.

INTRODUCTION

Today almost anyone can have a private voice conversation with almost anyone else anywhere. Developments in wireless technologies have extended the voice connectivity from stationary sites to people on the move. Phones are locally connected via wire or wireless paths to a wide area network. Wireless phone access is more expensive than wired service now, but recent spectrum allocations for Personal Communications Services (PCS) will lower this premium.

Private video and data services however are very expensive and tightly controlled, so that hardly anyone has wired access, and still fewer have wireless access to broadband services. With NII, our present expectation of private voice and public images will transition to total public and private communications, where individuals have access to voice, image, and information services that today are only being imagined. The promise of NII lies in three synergistic forces: the availability of wired bandwidth brought on by developments in fiber optics and signaling; the availability of low cost computing brought on by semiconductor advances; and the emergence of competition and choice brought on by new telecommunications policies worldwide. However, unless adequate spectrum is allocated, the full potential of the NII will not be realized.

This paper is organized into three parts. The first presents some possible applications of voice, graphic and video services for people on the move. The next section gives estimates of the spectrum required to support these services. The last section considers the role of government in supporting the Wireless NII initiative.

WIRELESS COMMUNICATIONS APPLICATIONS

Land mobile radio is used by all segments of the industrial, business, public safety, public service and land transportation mobile work force. Motivated by the need of the private sector to improve productivity and service, users will migrate to the specific communications solutions which provide the greatest advantage to their operations. By the year 2000, many services will be possible with land mobile radio. These services can be organized into two groups: those available now, and new services. Dispatch, phone, transaction processing, and limited decision processing are in place today and serve a growing user base each year. Full decision processing, facsimile, snapshot, slow video and full video represent new services for the wireless user. Each of these is described in the following paragraphs.

Dispatch

Radio dispatch service was the first two-way application of wireless communications for people on the move. It is characterized by one-to-many voice communications through a central dispatcher to a more than one person in the field. It is most efficient when field people are working cooperatively to accomplish a task, so that the broadcast message is useful to all. For example, if a group of 20 police officers are providing public safety services to a city, knowledge of city-wide activity is useful to all to optimize their efficiency and service level.

Phone

The extension of wired phone service to wireless communications began about 30 years ago. Phone service is characterized by one-to-one voice communications using a duplex channel so that speakers may interrupt each other.

Transaction Processing

Transaction processing is a communications category which accesses and updates data base information in a central location. Messages are brief and response time must be short. Wireless examples include rental car check-in, where a contract number is used to access the rental profile, and mileage and time information are added to the data base for billing.

Decision Processing

Decision processing differs from transaction processing in the amount of data being handled. Decision processing involves much higher information content than transaction processing, and usually involves explicit data manipulation by the end user. For example an insurance salesman would combine information from a potential customer with data base information to develop a price quote. If the quote is accepted, the salesman could execute the policy, bill the customer, and place the policy in effect immediately. This service type is the extension of current office computing environments to the mobile worker.

Facsimile

Facsimile service is an extension of this ubiquitous wireline capability to the field. The information is limited to a few pages, and is a black-and-white image of text or line drawings. The resolution of the received image is low, and may be displayed on paper or a video terminal.

Snapshot

The snapshot is a still picture which extends the facsimile image to high resolution, gray scale, or color. Examples include police mug shots, accident scene pictures, or marketing shots of real estate properties. The image may also be non-visual, such as a field x-ray scan of an accident victim.

Slow Rate Video

This communications method provides high resolution color images at rates of about one frame per second. Emergency medical response teams could provide video images of accident victims to doctors for immediate attention. High quality pictures are needed, but low frame rates are acceptable, since the scene is relatively stationary.

Full Rate Video

This service is similar to slow rate video, but would support full motion at rates of 30 frames per second or higher. Firemen could use this service to coordinate their efforts in fighting a large fire from several approaches.

SPECTRUM REQUIREMENTS

By the year 2000, major metropolitan areas in the US, such as New York, Chicago and Los Angeles, will have populations of about 18 million people each. Based on census data, about 3.7 million of these people will be moderately to highly mobile workers, spending 30 percent or higher of their work day away from the office. The needs of this group must now be estimated. The proposed usage model is shown in Table 1.

Service usage is expressed as number of requests per hour, average duration per request, and percentage usage by the mobile workforce. From this usage model and a technology model of the bandwidth required to support each service type, the total spectrum is also indicated. The spectrum per user assumes the continued development of efficient information source coding algorithms, modulation techniques and frequency re-use plans. The spectrum required is the product of the assumed user base, request rate, duration, percent usage by the mobile population, and spectrum per user.

Table 1. Year 2000 Service Usage Profile and Spectrum.

Service	Request Rate	Duration	Usage	kHz/User	Spectrum
Dispatch	1 per hr	40 sec	25%	5	51 MHz
Phone	1 per hr	140 sec	50%	2.5	180 MHz
Transaction Processing	2 per hr	2 sec	25%	5	5 MHz
Facsimile	0.5 per hr	40 sec	12%	10	25 MHz
Snapshot	0.2 per hr	60 sec	6%	10	7 MHz
Decision Processing	0.2 per hr	60 sec	12%	20	30 MHz
Slow Video	0.1 per hr	210 sec	2%	20	9 MHz
Full Video	0.1 per hr	210 sec	0.1%	300	6 MHz
TOTAL					**313 MHz**

Beyond the year 2000, technology developments and market acceptance will cause the spectral efficiency to improve and cost of service to decrease, leading to higher levels of penetration. In particular, the services beyond dispatch and phone will show significant growth. Table 2 shows an estimate of the expected usage profile, spectrum per user and resulting spectrum required in the year 2010.

Table 2. Year 2010 Service Usage Profile and Spectrum.

Service	Request Rate	Duration	Usage	kHz/User	Spectrum
Dispatch	1 per hr	40 sec	25%	5	51 MHz
Phone	1 per hr	140 sec	50%	2.5	180 MHz
Transaction Processing	2 per hr	2 sec	50%	5	10 MHz
Facsimile	0.5 per hr	40 sec	50%	5	51 MHz
Snapshot	0.2 per hr	60 sec	50%	5	31 MHz
Decision Processing	0.2 per hr	60 sec	50%	5	31 MHz
Slow Video	0.1 per hr	210 sec	20%	5	22 MHz
Full Video	0.1 per hr	210 sec	1%	150	32 MHz
TOTAL					**408 MHz**

THE ROLE OF GOVERNMENT

At the national level, government must take an active role in assuring that the vision of wireless NII becomes a reality. Wireless communications must become an explicit part of the NII dialog, and spectrum must be available to support the birth and growth of innovative wireless services and products. The government allocates spectrum through two separate processes. The first process divides the total spectrum into those bands intended for government use, and those bands for all other uses. The Federal Communications Commission (FCC) is responsible for all non-government spectrum bands. The FCC has always been responsive to the needs of industry and public safety to ensure continued leadership of the U.S. in developing and implementing new wireless services. This response has come in several forms. To create the 800 MHz Land Mobile and Cellular bands, TV channels were reallocated from existing spectrum. To create the Enhanced Specialized Mobile Radio service (ESMR), rules covering existing 800 MHz operation were relaxed. To create Low Earth Orbit satellite service, cooperative allocations with nations around the world were negotiated. In each case the FCC showed the vision needed to enable new classes of wireless service which benefit the productivity and safety of the country. The spectrum needed to enable the wireless NII will exceed that available within the FCC domain, and action is required from other agencies. Most government spectrum use is for the Department of Defense; with new technologies, some reallocation to commercial and public safety use is possible without compromising the efficiency or safety of current services. Congress must act to see that the best national interest is considered when these reallocations are made.

CONCLUSION

The NII will expand public and private access to information of all forms. This access must include wireless communications so that the full potential for improved productivity and safety are realized. Government support is essential to provide the vision of wireless access everywhere, and to make available spectrum to enable this vision. The U.S. has led the world with communications and computing visions in the past, and with mobility as part of the NII agenda it will do so again well into the next century.

STATUS OF FERRITE TECHNOLOGY FOR HIGH VOLUME MICROWAVE APPLICATIONS

Denis C. Webb
Naval Research Laboratory
Washington, DC 20375-5320

ABSTRACT

With the emergence of high volume commercial and military applications, there is a growing need to reduce the size and cost of microwave ferrite components, especially ferrite circulators, to be more compatible with monolithic integrated circuits. The Ferrite Development Consortium, consisting of leading US ferrite government, university and industrial institutions, was formed under Advanced Research Project Agency (ARPA) sponsorship to address these needs. Areas of Consortium technical activity include bulk and thick-film techniques for batch processing of ferrite devices, improved computer-aided-design tools and prototype demonstrations. This paper will review the Consortium's materials development needs and progress.

INTRODUCTION

Ferrite components have played an important role in microwave systems for nearly forty years.[1] Their inherent capability of meeting demanding loss and power-handling requirements make them the technology of choice for numerous applications especially at high microwave and millimeter wave frequencies. There are an increasing number of applications, however, which place a premium on size and cost. Examples include military systems such as active arrays, decoys and missile seekers and commercial systems such as wireless communications and automobile radars. In these systems semiconductor technology, and especially monolithic-microwave-integrated-circuits (MMICs) offers good performance and very low size and cost. There are several important microwave components, however, which are based on the unique characteristics of ferrites, e.g. circulators, tunable oscillators and filters, and frequency selective limiters which can not be readily duplicated by competitive technologies. There is therefore a growing need to implement these components in compact low-cost configurations which are compatible with semiconductor and in particular with MMIC technology.

Integrated ferrite components were first investigated in the late 1950's [2] but it has been only over the past few years that their development has been pursued aggressively. This activity has been spurred on not only by growing system needs but by advances in materials technology which allow novel device geometries to be realized. Much of the recent work has been carried out under the ARPA Ferrite Development Consortium, with the goal of achieving order-of-magnitude reduction in the size and cost of MMIC-compatible ferrite components, principally circulators. One of the principal elements of the program is development of techniques to produce low-cost ferrites in film and bulk form for high volume applications. Most of the results cited in this paper are derived from work performed by Consortium members.

MATERIALS REQUIREMENTS

This paper will primarily address the design and fabrication of the 3-port circulator since it is the most widely used of the integrated ferrite components. It is shown schematically in Figure 1. Power incident upon port 1 is transmitted to port 2 with little loss and minimal power transferred to port 3. Power incident upon port 2 appears at port 3, etc. The lowest-cost, most compact configuration is based upon microstrip circuitry with the junction area defined by a metallic shield on the ferrite (see Figure 2). The shield is commonly circular in shape but any shape with three fold symmetry can be used. The junction is biased with a static field perpendicular to the plane of the shield such that the ferrite is just completely saturated. One or more transmission-line matching sections (e.g. quarter-wave transformers) are commonly used to transform the circulator input-impedance to that of the rest of the circuitry.

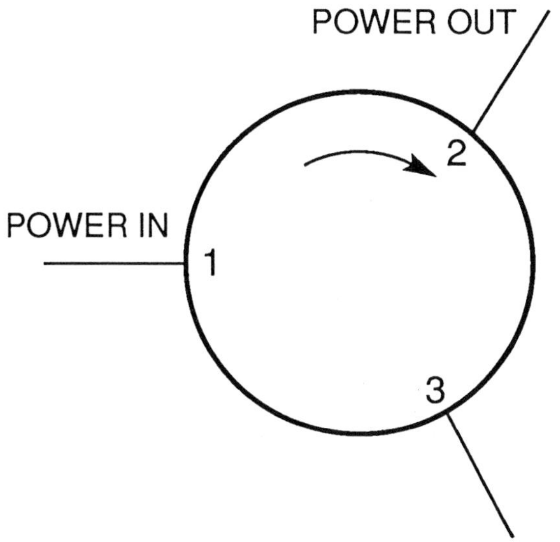

Figure 1. Junction circulator schematic

Several different generic ferrite topologies are used for circulators. The simplest employs a metal shield on an all-ferrite substrate. Its principal drawback is that the matching sections are on a ferrite material which increases loss and complicates design. A more common approach is to imbed a ferrite puck in a low-loss dielectric material such as alumina. This circumvents the above cited matching transformer problem but complicates fabrication. The ferrite must be firmly attached to the dielectric with a minimal discontinuity between the materials for all required temperature extremes. For ultra broadband applications, e.g. 6-18 GHz, it is advantageous to use more than one ferrite in the junction area to produce a more uniform bias field within the active region.[3] For a circular design, a high saturation magnetization ferrite puck is surrounded by a concentric ring of a lower magnetization ferrite. This combination is then embedded in a low loss dielectric.

As noted above, many of the emerging high volume applications place a premium on size. There are naturally some basic physical as well as practical constraints which limit the size reduction that is possible. For example, the distributed junction-type circulator shown in Figure 2 requires

Figure 2. Ferrite Y-junction microstrip circulator (courtesy of Electromagnetic Sciences).

a specific mode pattern in the junction area to achieve efficient circulation. The desired mode has maxima at the excitation and coupled ports and a null at the isolated port. This dictates that the shield diameter be approximately one half electromagnetic wavelength in the ferrite medium, i.e. ~ l/2 cm at 10 GHz. Lumped-element configurations have been investigated to circumvent this size limitation, but research has primarily addressed low microwave frequency, narrow-band applications where loss is moderate. Although recent work indicates that these concepts may prove useful above 5 GHz,[4] they will still be unable to address a large number of circulator needs.

In reducing the thickness of the ferrite material there are no fundamental limits dictated by the modal structure since the RF fields do not vary in the thickness direction. However, there are important design tradeoffs between loss, impedance and thickness. These are illustrated by the design example shown in Figure 3 for a circulator designed to span the 3-10 GHz range. The calculation is based on a classical Green's function approach modified to include dielectric, magnetic and metal losses.[5] Note that at thicknesses exceeding .020", the loss is relatively independent of thickness and is dominated by dielectric and magnetic contributions. At thicknesses below .010" metal loss begins to dominate and increases rapidly with decreasing thickness. Furthermore, the circulator impedance decreases very rapidly in this range, increasing the losses associated with matching. Thus for this particular example, it is not practical to employ thicknesses much less than .010". The relationship between thickness and performance is naturally dependent on electrical design and available materials but qualitatively there is always a point where reducing thickness leads to rapidly deteriorating performance.

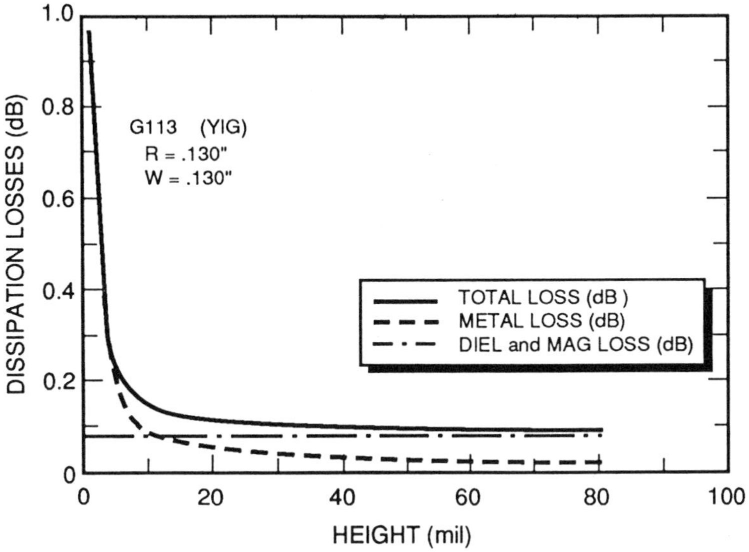

Figure 3. Dissipation losses of a C-band microstrip circulator

Finally, the overall size of the circulator is commonly dominated by the magnet and circuitry. At millimeter-wave frequencies, the increased field required and smaller circuitry makes the problem especially severe. Fortunately, there are classes of ferrites which are capable of alleviating this problem by providing their own internal bias and thus require little or no external bias. Initial circulators have been demonstrated at millimeter wavelengths with these materials.[6] Concepts to extend operation to lower frequencies are being pursued in the Consortium.

Another promising approach for reducing magnet size is the deposition of rare earth permanent film-magnets directly into the microwave structure. Sm-Co films of energy products of 20 kg-Oe and thicknesses of 100 μm have been demonstrated. When deposited directly on YIG substrates these films have been shown to be capable of biasing the YIG to saturation.[7]

A large variety of ferrites are used for microwave applications. At low to mid microwave frequencies garnets (Gd-YIG, YIG) and spinel ferrites (MgMn, Li TiMn) with saturation magnetization in the range 1000-2500 gauss are commonly used. High microwave and millimeter wave frequency circulators employ spinels with a saturation magnetization of 3000 gauss and higher. Self-biased circulators typically use Ba M - type hexagonal ferrites. All materials must exhibit a small loss tangent ($\leq .001$) and spin-wave linewidth ($\Delta H_K \leq 10$ Oe) to minimize dielectric and magnetic losses, respectively. Finally, low sensitivity to temperature variations is often important.

BULK MATERIALS

The ARPA Ferrite Development Consortium is placing its primary materials emphasis on films; however, there are important ongoing efforts in bulk ferrite growth, especially for high volume applications. It is evident from the previous section that low-frequency and wide-bandwidth applications do not lend themselves well to film technology because excessive thickness is required. Thin bulk samples can also provide a transition mechanism for the less well-developed film approaches and serve as standards for judging film quality. This section will describe recent progress in preparing bulk ferrite materials to enhance the manufacturability of microstrip junction circulators.

Ferrites have been traditionally made by sintering component oxides at temperatures between 1000°C and 1450°C.[8] Preparation is a complex process entailing milling of raw materials to produce a homogenous mixture of fine particles, pre-sintering and remilling to minimize subsequent shrinkage and then sintering again to produce the needed chemical reaction and crystallization. This is followed by slow cooling and grinding to the desired shape.

Tape casting is an intrinsically economical technique for producing large thin sheets of ceramic materials such as are required for microstrip circulators. Although tape casting of ceramics has been employed for many years, efforts have been almost exclusively focused on obtaining small-grained, electrically isotropic materials. The general tape casting approach[9] entails forming a slurry of the desired material with an organic binder and solvent and then using a doctor blade to spread the material to a uniform thickness to produce green sheets. The sheets are heated to evaporate the binder and sintered at high temperatures. Under Consortium sponsorship, Trans-Tech has applied this technique to yttrium-iron-garnet. Fully-dense tapes have been produced with magnetic and dielectric properties equivalent to conventional bulk ceramic counterparts. Work is underway to improve surface finish so that a final machining step is not required.

Closely related to tape casting is roll-compaction. Here, the material is drawn through rollers rather than being thinned via a blade as is done in tape casting. Again under Consortium sponsorship Trans-Tech has applied this technique to both yttrium-iron-garnet and lithium ferrite. Fully-dense materials with excellent electrical and magnetic properties were achieved with this technique.

It was noted in the previous section that multiple ferrites are sometimes required for very wideband applications. The standard fabrication approach used for this configuration is to adhesively bond the different materials together. This poses manufacturability problems since it is difficult to achieve a reproducible well-controlled junction between the materials. A superior approach is to co-fire the high magnetization plug insert in a lower magnetization substrate. Successful carrying out of this process mandates good dimensional control under wide temperature variations to maintain intimate contact between the materials without undue stress. Raytheon has successfully demonstrated this technique using lithium manganese-titanium ferrites ($4\pi M_{S1}$ = 2750 gauss, $4\pi M_{S2}$ = 1300 gauss). The bond between the ferrites was excellent and properties were indistinguishable from separately formed samples. Work is in progress to incorporate this material combination into a circulator.

FILM REQUIREMENTS

Since the early 1970's, the application of magnetic films to microwave devices has been under investigation. Until quite recently, work has been largely concentrated on devices based upon single crystal YIG films grown epitaxially on gadolinium-gallium-garnet (GGG) substrates. The principal goal of these activities has been to achieve a compact low-cost replacement of the classical tunable YIG-sphere oscillators and filters.[10] The combination of high-Q and wide tunability of YIG devices is unsurpassed by any other technology. Fixed-tuned YIG film devices have also proven to be a competitive technology at low microwave frequencies.[11] Since they are based on ferromagnetic resonance phenomena, size is not directly dependent upon the electromagnetic wavelength.

It was not until the late 1980's that researchers began to seriously consider applying ferrite film technology to mass-producible MMIC compatible circulators. Early work performed by Abe and co-workers at Tokyo Institute of Technology demonstrated that spinel ferrites could be grown on a variety of substrates, including GaAs, by electroless plating at temperatures below 100°C.[12] Subsequently a number of other techniques have been studied to address this application; those described in the following Section are being actively pursued by members of the Consortium. Before discussing these techniques, general film requirements for microstrip circulators will be outlined.

Achieving thick, mechanically stable ferrite films with good magnetic and electrical properties on a suitable microwave substrate is a significant technical challenge. High firing temperatures and thermal mismatch between ferrite and substrate compound the problem. The coefficient of expansion for typical ferrites and substrates of interest for monolithic devices is shown in Table I. Note the higher values of CTE of the ferrites as compared to the semiconductor substrates. Table II indicates material and deposition parameter goals for X-band and millimeter-wave

Table I.
Coefficient of Thermal Expansion

Material	CTE(X100^{-6}/°C)
Si	3.6
GaAs	6.0
Al_2O_3	8.0
$Y_3Fe_5O_{12}$	10.4
$MgMnZnFe_2O_4$	8.2
$BaFe_{12}O_{19}$	7.0

Table II.
Film Characteristics for Microwave and Millimeter Wave Circulators

	Spinels and Garnets; X-band circulator Goal	Hexagonals; MMW circulators Goal
<u>Physical Properties</u>		
Thickness	100 μm	50 μm
Area	50 cm^2	50 cm^2
Surface Finish	10 micro-in	5 micro-in
Surface adhesion tape test	passes 100% of time	passes 100% of time
Crack-free growth	no cracks	no cracks
<u>Process Parameters</u>		
Deposition rate	2500μm cm^2hr^{-1}	2500μm cm^2hr^{-1}
Deposition and extended anneal temperature	700°C	700°C
<u>Material Parameters</u>		
Saturation magnetization	2000-2500 gauss	>4000 gauss
Magnetic linewidth	200 Oe	500 Oe
Loss tangent	.001	.001
Coercive field	---	50% $4\pi Ms$
Remanent Magentization	---	90% $4\pi Ms$

circulators. Area and deposition rate (volume of material per unit time) goals are cited to enhance commercial viability. Good surface morphology is necessary to minimize scattering loss and allow for subsequent metallization and patterning. Magnetic linewidth, loss tangent and thickness goals are commensurate with low loss for moderate bandwidths. The millimeter wave device is self-biasing hence uses a hexagonal ferrite. High coercive field and remnant magnetization are required for stable biasing.

FILM DEPOSITION TECHNIQUES

Spin-Spray

It was noted above that the first technique investigated for producing MMIC-compatible circulators was electroless plating. The most successful of the electroless techniques was spin-spray. In spin-spray a chloride solution of the spinel transition metal constituents (Fe, Ni, Zn etc.) and a separate oxidizing solution are sprayed sequentially onto a heated spinning substrate. The film is built up layer-by-layer by successive deposition and oxidation steps. The principal advantage of this technique is the low deposition temperature, typically 100°C to 600°C, which makes it compatible with processing of GaAs MMICs. Films grown by this approach have exhibited excellent magnetic properties, e.g. a linewidth of less than 32 oersteds has been achieved for a Ni-Zn ferrite at 9.3 GHz. Growth rate is moderate, typically 2 μm/hour but the process could in principle be scaled up to large areas, thus achieving acceptable throughput. The principal problem encountered with this technique is in taking the oxidation reaction to completion ($Fe^{+2} \rightarrow Fe^{+3}$), particularly for thick films. Small amounts of divalent iron present in the final material give rise to excessive electrical loss. Tan δ for spin-spray films produced by the Consortium has been in the range .1 to 1 at 10 GHz, unacceptably large for circulator applications. Attempts to replace the divalent iron with other divalent metal ions by varying the solution constituents and anneal cycles have not yet been successful. Resolution of this problem is required before spin-spray can be used for microwave applications.

Pulsed-Laser-Deposition

Pulsed-laser-deposition (PLD) has proven to be an excellent technique for producing high quality films of chemical compounds. It has been successfully applied to a variety of metal oxides including high-temperature superconductors, ferroelectrics, dielectrics and ferrites. PLD is a conceptually simple flash evaporation technique which transfers the stoichiometry of the target material to the film. Chemical interactions at the deposited surface can result in epitaxial growth, hence films with excellent magnetic and electrical properties. Growth rates are moderate, typically 1 μm to 30 μm per hour depending upon substrate size and growth parameters. High quality spinel, garnet and hexagonal ferrite thin films have all been produced by PLD. For example, Mn-Zn ferrites with in-plane linewidths as low as 40 oersteds at microwave frequencies have been achieved.[13] Barium-ferrite has been grown epitaxially on a sapphire substrate with a resonance linewidth of only 66 oersteds at 58 GHz.[14] Films up to 20 μm in thickness have exhibited good magnetic properties but growth of high quality thicker films has not been possible due to CTE mismatch problems. Thin yttrium-iron-garnet films produced by PLD have achieved linewidths approaching 1 oersted when grown epitaxially on lattice-matched GGG.[15] Thick (75 μm) films grown on metallized GaAs have exhibited bulk values of magnetization and magnetic linewidths less than 50 Oe, more than acceptable for microwave applications.

Improvements needed to address the circulator application are better mechanical properties and loss tangent. Tan δ is ~ .01, an-order-of magnitude higher than desired. YIG films in the 50 to 100 μm thickness range have been produced on silicon and gallium arsenide substrates but the large CTE mismatch between ferrite and substrate has led to high strains and cracking of the films. A YIG circulator with 20 dB of isolation was successfully fabricated on silicon by PLD but the thermally induced cracks in the YIG film resulted in high (8 dB) insertion loss.[16]

Pulsed-laser-deposition has been primarily a laboratory tool which allows researchers to deposit films of complex materials that are difficult to produce by other physical deposition processes. Recently, however, the process has been scaled up to 150 mm diameter wafers for a variety of metal-oxides including yttrium-iron-garnet.[17] A 1 μm YIG film had less than a six percent variation in thickness over the central 125 mm substrate diameter. Energy dispersive X-ray analyses of the films showed that the film composition basically remained constant during the entire deposition process.

Thus, PLD is a promising technique for high-volume production of circulators. The principal problem areas, thermally induced cracking and high electrical loss, are being addressed through the use of intermediate layers and optimization of growth and anneal conditions.

Cyclic-Coat-and-Fire

NIST is developing a cyclic-coat-and-fire technique under ARPA Consortium sponsorship for growing thick ferrite films.[18] It is a continuously operating process in which a substrate is sequentially dipped into an aqueous metal nitrate solution and fired at a temperature ranging from 550°C to 1100°C depending upon the ferrite. The sample is then returned to the solution and the cycle is repeated. Five to ten nm of material are deposited during each cycle. Deposition rates are slow compared to most other film growth techniques but because it can be readily automated to run continuously and scaled to large areas it lends itself well to manufacturability. Cyclic-coat-and-fire has been used to produce both garnet and spinel films on magnesium-aluminum-titanate. Adhesion of these films has been excellent and surface finish good (~20 μm inches). Samples of up to 50 μm in thickness have been produced. Linewidths are comparable to bulk values but loss tangent is moderately high (~.005) and saturation magnetization is only one half to two thirds of the bulk values.

Jet-Vapor Deposition

The jet-vapor process is a high speed deposition process which uses sonic jets of inert materials in a low vacuum.[19] Helium, argon or hydrogen jets are formed by a nozzle and propel vaporized materials to substrates at high speed (1 Km/sec). The substrates are in a low-pressure, fast-flow system which is maintained in steady-state by a high-speed (20,000 ℓ/min) mechanical pump. The vaporized metal is generated within the nozzle by several methods, including evaporation, laser ablation and plasma chemical reactions. Films of alloys or compounds can be deposited by operating more than one jet simultaneously. As many as four jet sources have been used to produce multicomponent device quality films such as PZT [Pb(Zr, Ti) O_3]. The process is readily scaled to high volume manufacturing by mounting samples on a drum-shaped rotor perpendicular to the jet axis which can be spun up to rates of 100 H_Z. The motor is capable of moving up and down while the samples are spun horizontally thus assuring every point on the substrate is exposed for an equal time to the jet.

Principal advantages of the approach are simplicity, versatility and high throughput. In the ARPA Consortium Jet Process Corporation has concentrated on growth of nickel ferrite films. Films more than 125 μm in thickness have been produced at rates of 2-5 μm/minute. Problems encountered with the process are similar to those encountered by other techniques, namely, high loss tangent (~ .01) and low magnetization (50-60% of the bulk value) for the very thick films.

SUMMARY AND CONCLUSIONS

There has been considerable progress over the past two years in advancing ferrite materials technology for high volume microwave applications, especially for microstrip circulators. Low frequencies, broad bandwidths and high power dictate bulk approaches. Tape casting and roll-compaction are particularly attractive for fulfilling these needs and considerable progress has been achieved in applying these to microwave ferrites.

Film technology is being pursued for higher frequencies and moderate fractional bandwidths. Millimeter wave applications are especially attractive since thickness requirements are less demanding and geometries lend themselves to photolithograpic techniques for mass production. Hexagonal ferrites are being investigated for use at millimeter wavelengths to eliminate the need for the principal volume and weight contributor - the permanent magnet. Recent progress in producing quality thick (50 μm to 125 μm) ferrite films has been excellent. Work is now concentrating on reducing dielectric loss, raising saturation magnetization and improving mechanical properties to better approach bulk values. Initial film circulators have already been demonstrated; prognosis for achieving performance comparable to bulk counterparts at a fraction of the cost and volume is excellent.

ACKNOWLEDGMENTS

The author is indebted to ARPA Consortium members from Westinghouse, Trans-Tech, Raytheon, EMS for providing material for this paper, especially Douglas Adam, Ernst Schloemann, John Simmins and Tom Taylor. Financial support and encouragement from Frank Patten, Stu Wolf and Lee Buchanan of ARPA is also acknowledged.

REFERENCES

1. L. R. Whicker, Ferrite Control Components, Vol. 1, Junction Circulators, YIG Filters and Limiters, Artech House, Inc., Dedham, MA, 1974.
2. G.T. Roone and H. A. Hair, "Thin ferrite devices for microwave integrated circuits," IEEE Trans. on MTT, MTT-16, No. 7, pp. 275-284, July 1968.
3. R.E Blight and E. Schloemann, "A compact broadband circulator for phased array antenna modules," 1992 IEEE MTT-S Intl. Microwave Symposium Digest, pp. 1309-1392.
4. Ch. Ngoonou Kovam, J.P. Coupez, S. Toutain, G. Forterre, P. Desmarest, "Frequency limits of the ultra-miniature circulator," 1994 IEEE MTT-S Intl. Microwave Symposium Digest, pp. 929-931, May 1994
5. R. E. Neidert and P.M. Phillips, " "Losses in Y-junction stripline and microstrip ferrite circulators," IEEE Trans. on Microwave Theory and Techniques, pp. 1081-1086, July 1993.
6. J.A. Weiss, N. G. Watson and G. F. Dionne, "Now uniaxial-ferrite millimeter wave junction circulators," 1989 MTT-S Intl. Microwave Symposium Digest, pp. 145-148, June 1989.
7. F.J. Cadieu, H. Hejde, E. Schloemann, H. J. Van Hook, "In-plane magnetized YIG substrates self-biased by SmCo based sputtered film coatings," 6th Joint MMM-INTERMAG, Albuquerque, NM, June 1994.

8. P. J .B. Clarricoats, Microwave Ferrites, John Wiley and Sons Inc., New York, 1961.
9. M. Ishu, Maeda, M. Takata and T. Yamashita, "Fabrication of superconducting YBCO films by a tape casting method," Jap. Journal of Appl. Phys.. Vol. 26, No 12, L1959-L1960, Dec. 1987.
10. W. S. Ishak, "Magnetostatic wave technology; A review," Proc. IEEE, Vol. 76, No 2, pp. 171-187, Feb. 1988.
11. Y. Murakami, "Microwave ferrite technology in Japan: Current status and future expectations," 1993 MTT-S Intl. Microwave Symposium Digest, pp. 207-210, June 1993.
12. M. Abe, T. Itoh, Y. Tamaura, Y. Gotoh and M. Gomi, "Ferrite plating on GaAs for microwave monolithic integrated circuit," IEEE Trans. on Magnetics, 23, No. 5 Sept. 1987, 3736-3738.
13. C. M. Williams, D. B. Chrisey, P. Lubitz, K. S. Grabowski and C. M. Cotell, "The magnetic and structural properties of pulsed laser deposited epitaxial MnZn ferrite films," J. Appl.. Phys. 75 (3), pp. 1676-1680, Feb. 1994.
14. C. A. Carosella, D. B. Chrisey, P. Lubitz, J. S. Horwitz P. Dorsey, R. Seed and C. Vittoria, "Pulsed laser deposition of epitaxial BaFe12O19 thin films," J. Appl.. Phys.. Vol. 71 No. 10, pp. 5107-5110, May 1992.
15. P. C. Dorsey, S. E. Bushell, R. G. Seed and C. Vittoria, "Epitaxial yttrium iron garnet films grown by pulsed laser deposition," J. Appl.. Phys.., Vol. 74, No. 2, pp. 1242-1245, July 1993.
16. J. D. Adam, H. Buhay, M. R. Daniel, M. C. Driver, G. W. Eldridge, M. H. Hanes, and R.L. Messham, "Monolithic integration of an X-band circulator with GaAs MMICs", 1994 MTT-S Symposium, Orlando, FL, May 1994.
17. J. Greer, and M. Tabat, "Large-area pulsed laser deposition; techniques and applications," American Vacuum Society Meeting, Denver, CO, Oct. 1994.
18. J. Ritter, NIST private communication.
19. B. L. Halpern, J. J. Schmitt, J. W. Golz, Y. Di and D. L. Johnson, "Gas jet deposition of thin films," Applied Surface Science, Vol. 48/49, pp. 19-26, 1991.

CERAMIC TECHNOLOGIES FOR PORTABLE RADIO APPLICATIONS

Wei-Yean Howng, Ceramic Technologies Research Lab., Motorola, 4800 Alameda Blvd., NE, Albuquerque, NM 87111

ABSTRACT

Ceramic technologies have become the essence of modern portable radio communicators. The use of low loss, high dielectric constant ceramics for RF and IF filtering, voltage controlled oscillators, and capacitors has enabled us to miniaturize the portable radio to a practical size while maintaining high performance. In response to the continuous quest for a smaller and thinner profile, ceramic devices are transforming from discrete components into monolithic lumped circuit component and finally into a totally integrated substrate. Resonators, capacitors, inductors, and resistors are being integrated into a high density multilayer structure. Compared to the organic circuit board, the multilayer ceramic substrate offers advantages such as lower processing cost, lower RF insertion loss, better interference isolation, higher component density and better temperature properties.

INTRODUCTION

Modern civilization was first enabled by the advent of power devices such as the steam engine and the gasoline engine that greatly enhanced human mobility. These power devices allow us to reach longer distances and cover more area. The consequence was a tremendous cycle time reduction. A person now achieves a hundred-fold increase in life time accomplishments, compared to his ancestors.

The second stage of modern civilization was marked by the advent of new electrical and electronic technologies which provide in-house conveniences such as TV and VCR. Unconsciously, these conveniences led to self-induced confinement in the home. Although two-way radios have been available for more than 50 years, the real liberating impact for human beings did not occur until the introduction of cordless phones, cellular phones, and pagers during the last ten years.

However, the popularity of the wireless communicator is far from ideal because of cost and portability limitations. The size and weight of the radio are still dictated by the dimension of the passive components and the battery. In fact, semiconductor integration shrunk the circuit board so much that discrete passive components looked like space thieves. Therefore, new ceramic technologies are desperately needed to alleviate the embarrassment.

Fig.1 The portable phone liberated human beings from the confinement caused by the TV, VCR, etc., electronic technologies.

NEW CERAMIC TECHNOLOGIES

Generally speaking, every passive component behaves like a distributed circuit at RF frequency. The parasitic effect, in many cases, overwhelms the component's original intent. The equivalent circuits of the passive components are shown in Fig. 2. As frequency increases, the parasitic resistance (skin effect) and the parasitic reactance become so appreciable that an impedance peak can be produced at or near the application frequency, thereby making the electrical properties uncontrollable. This problem is more pronounced at frequencies greater than 500 MHz. Recent ceramic technologies enable us to fabricate the component in chip format and greatly reduce the unwanted lead wire's inductance, distributed parasitic capacitance, and reduce the temperature variation effect.

It is well known that titanate based ferroelectric ceramics or the relaxor dielectrics with palladium electrodes, achieve high capacitance in very small size in a multilayer form [1]. The size reduction is readily appreciated in the surface-mount format. However, it is still a thousand times larger than semiconductor integration, thereby leaving ceramic devices behind semiconductor integration.

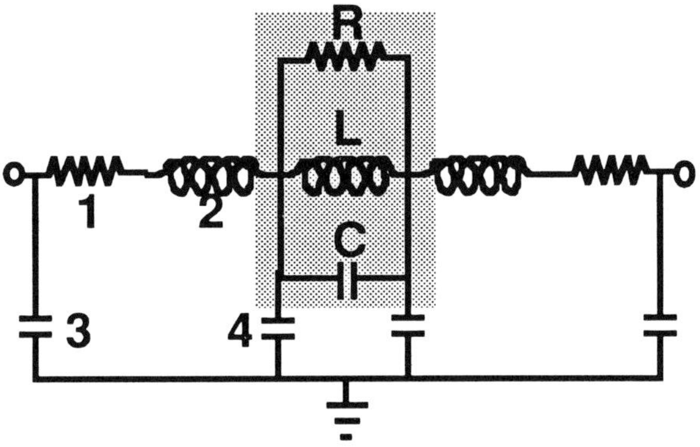

Fig. 2 The generalized equivalent circuit representing discrete components at high frequency. The component is a resistor, a capacitor, or an inductor depending on which—R, L, C—value is dominating. Element 1 is the lead resistance, 2 is the lead inductance, 3 and 4 are the parasitic capacitance when mounted near the ground plane. The impedance contributed by 1 and 2 become appreciable at high frequency.

Compared to capacitors, inductors are even worse. The inductor has had almost no improvement for many decades until the introduction of the multilayer inductor and thin film spiral inductor. The resistor has also migrated from leaded axial shape to the surface mountable chip form, which has greatly reduced the size—especially the height. All of these devices use ceramics as the carrier or as the functional body. The quality factors are still far from satisfactory for UHF and SHF applications.

The equivalent circuit self-resonant frequency of a discrete component has to be designed to be far away from the application frequency. Near the resonant frequency, the RF property is very unpredictable due to the process control limitation. The stray capacitance, induction, current leakage is minimized using proper ceramics and metallization scheme.

For wireless communication applications, the RF circuit, signal processing circuit, and the power supply are the representative circuits (see fig. 4). There are many particular components, such as the band pass filters, VCO, transmission lines that are not used in regular electrical circuits. These components have many unique requirements. The transmission line needs to have matched impedance that requires a specific geometric ratio and low resistivity conductor. The resonators and the filters have to have extremely low

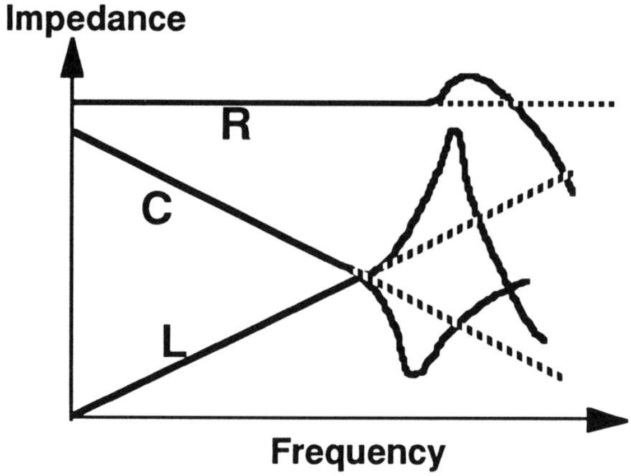

Fig. 3 Parasitic elements cause the impedance to have peaks at high frequency. Near the resonant frequency, the RF property is very unpredictable.

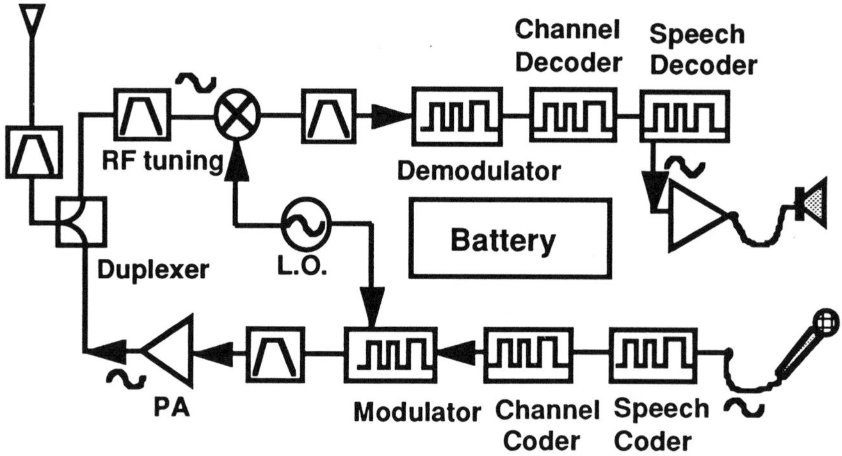

Fig. 4 Simplified two-way radio block diagram indicates the main ceramic filter components.

temperature coefficient of the resonant frequency. Unlike the traditional ceramic multilayer package, such as QFP emphasizing only fine pitch and line width, different design rules are applied. The insertion loss of all of the RF components has to be minimized to conserve energy. The insertion loss is mostly contributed by the conductor electrical resistance. Since the conductor line width is predetermined in order to match the RF impedance, the resistivity of the conductor has to be as low as possible. Thus, we have no other choice but to use copper or silver electrodes. Aluminum not only has a low melting temperature which is too low for co-firing but also has a resistance about 65% higher than silver or copper. Although gold is co-firable, the resistivity and cost are too high for commercial applications.

Using a high K dielectric, one reduces the co-axial resonator volume tremendously (fig. 5). For example, since the RF wave length in a media is proportional to the square-root of the dielectric constant, using the barium modified neodymium titanant with a dielectric constant of 100 will reduce the volume by a thousand times as compared to the air cavity. The co-axial resonator performs better than the equivalent lumped circuit with discrete components. Although higher dielectric constant ceramics can logically be used to reduce the size further, the insertion loss and manufacturing problems appear insurmountable.

Higher dielectric constant materials have two disadvantages at high frequency. They tend to have higher dielectric loss and temperature sensitivity. To maintain proper impedance, the center conductor becomes too narrow to be formed due to processing limitations. Narrow center electrodes also lead to Q degradation.

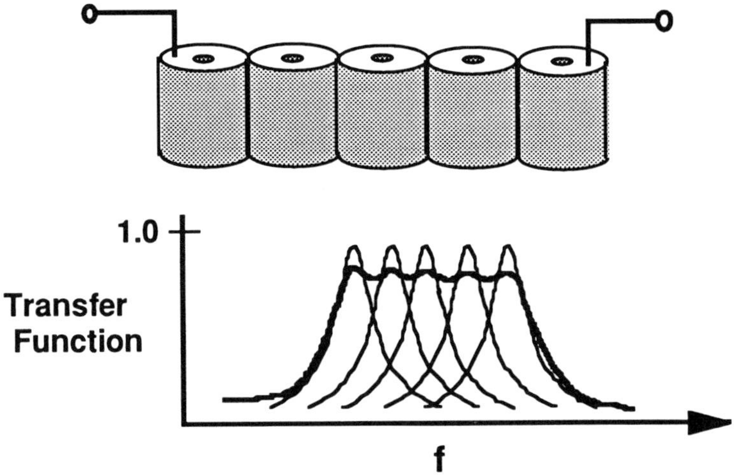

Fig. 5 The ceramic band-pass-filter can be fabricated with a collection of co-axial resonators. High dielectric constant ceramic is used to reduce the height of the quarter wave resonator. The insertion loss is contributed by the electrodes and the dielectric loss.

FUTURE TREND

It is becoming obvious that only the PCMCIA format or the so-called credit card size format provides true portability. The thickness therefore needs to be reduced to less than 2 mm. In order to reduce the thickness, the surface mount approach faces very tough challenges. Instead of mounting components on a printed wire board, the ceramic components themselves should also function as the board and IC chip carrier (see fig. 6).

All components are embedded into the substrate while recess cavities are provided for mounting active IC chips using the dropped in, the flip-chip technology. Since the resistance of the metal interconnects contribute to RF loss, refractory metals can not be used. Therefore, the ceramics need to be co-firable with either copper or silver. The sintering temperature should be less than 950 ^{o}C while maintaining high dielectric and magnetic Q factors. The structure will be like the exhibit in fig. 6. All the circuit and components are integrated into a 3-D structure with minimized thickness. For RF resonators and transmission lines, a virtual space is created using new design technology to provide the required geometry ratio for impedance matching (see fig. 7).

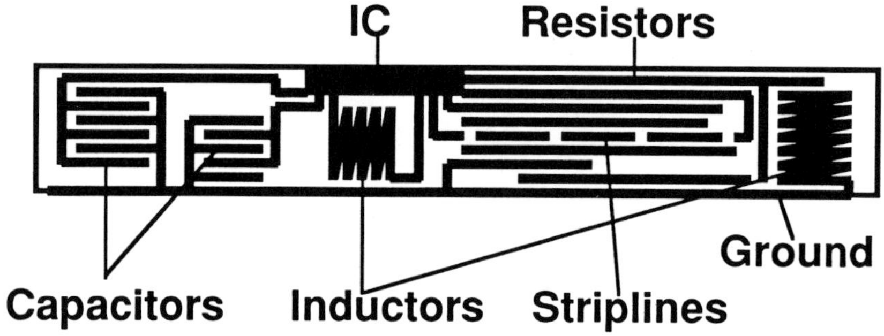

Fig. 6 All passive and active components are embedded into a multilayer substrate to achieve the thickness goal of the PCMCIA format.

CONCLUSION

To achieve true portability, new materials having different dielectric constant and permeability with low RF loss need to be developed in conjunction with the low resistivity metallization. However, before the room temperature superconductor is discovered, the burden lies solely on the shoulder of ceramic materials. The discrete surface mount approach can only squeeze the circuit to a certain extent. Most of the space is wasted for component terminations for soldering. The thickness can not be reduced effectively because of the need for wire board or carrier. An integrated multilayer approach embedding both the passive components and inter-connects as well as adding recessed cavities for flip-chip active IC mounting is the next logical way to realize the truly portable radio communicators for the next generation (fig. 7 and Fig. 8).

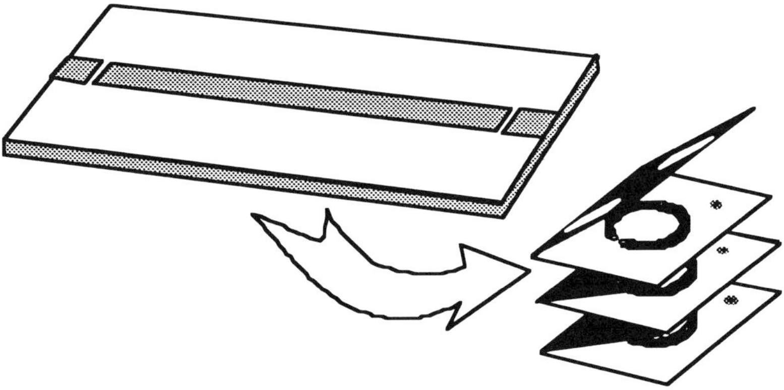

Fig. 7 Although finite dimension is required for RF components, a virtual space is realized using advanced ceramic technologies.

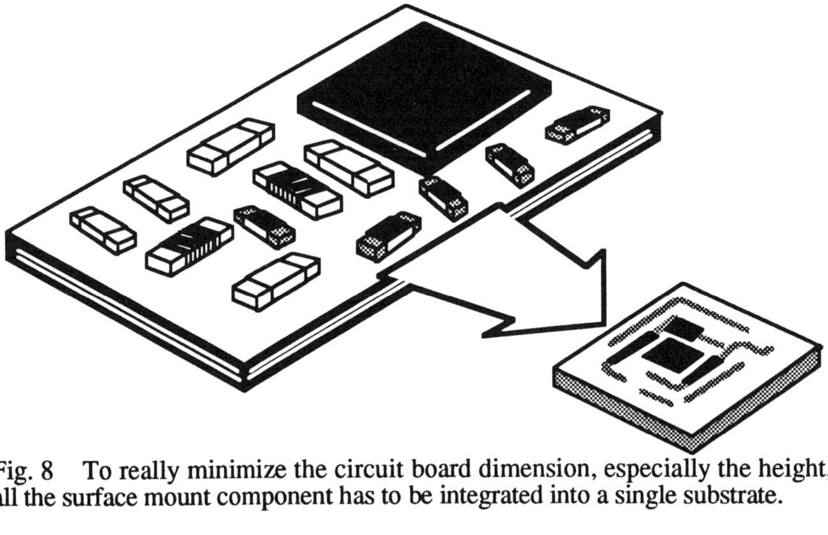

Fig. 8 To really minimize the circuit board dimension, especially the height, all the surface mount component has to be integrated into a single substrate.

REFERENCE

1. M. Kahn, D. P. Burks, I. Burn, and W. A. Schulze, "Ceramic Capacitor Technology", in Electronic Ceramics, L.M. Levinson, ed.,Marcel dekker, Inc., New York and Basel, 1988, pp. 191-274.

THE NIST METROLOGY PROGRAM ON ELECTROMAGNETIC CHARACTERIZATION OF MATERIALS*

Claude M. Weil
National Institute of Standards and Technology (813.08)
325 Broadway, Boulder, CO 80303-3328

ABSTRACT

The Electromagnetic Properties of Materials (EPM) Program at the National Institute of Standards and Technology (NIST) is described, including an outline of the current goals of the project and details of measurement techniques being used at NIST for characterizing dielectric and magnetic materials of importance in wireless communications in the RF spectrum of interest.

INTRODUCTION

There are many important applications of low-loss ceramic materials in high-Q components used in wireless communication technologies, such as filters, voltage controlled oscillators, tuned amplifiers, etc. Other applications include printed circuit/wiring boards and substrates. Ferrites find wide application in microwave isolators, circulators and phase-shifters. An essential prerequisite for the successful use of any of these materials are accurate and reliable data on the complex permittivity, $\epsilon^*=(\epsilon_r'-j\epsilon_r'')\epsilon_0$ and permeability, $\mu^*=(\mu_r'-j\mu_r'')\mu_0$ of such materials throughout those portions of the RF/microwave spectrum of interest in wireless communication. It is generally agreed that the available data are often inadequate and that the quality of the metrology used to generate these data needs improvement.

*Contribution of the National Institute of Standards and Technology, not subject to copyright.

This paper reviews the details of the materials characterization program at NIST

PROGRAM OBJECTIVES

The goals of the NIST program may be summarized as follows:

(a) Evaluate existing measurement methods, assess their accuracy, and improve on them where possible. Develop new measurement techniques and software as needed. Much of the NIST effort, to date, has concentrated on metrology improvements.

(b) Select materials that have very stable electromagnetic (EM) and mechanical properties and are suitable for use as standard reference materials (for example, rexolite, fused quartz, alumina). After these materials are accurately characterized by NIST, they are subsequently made available to industry for use as check standards for purposes of calibrating their own measurement systems and software. Such a system partially satisfies the requirements for traceability to NIST in material measurements.

(c) Provide NIST measurement or calibration services to those requiring characterization of their own materials. Provide consultation and assistance to industry and others on their measurement problems.

(d) Participate in national and international intercomparisons of material property measurements.

(e) Develop a NIST database on commonly used materials.

In keeping with these basic priorities, much of the NIST effort to date has concentrated on metrology improvements for characterizing homogeneous and isotropic solid materials in the frequency range from 300 kHz to 18 GHz; some measurements of liquid and granular dielectrics have also been undertaken. Most material measurements are performed at room temperature, but NIST does now have capabilities for performing cryogenic measurements. As the program expands in response to customer needs, both the frequency coverage and temperature range will be extended beyond their existing limits.

The reader is referred to previous reviews of the NIST program, published earlier.[1,2] This review will not attempt to repeat this material, but rather to update it, with emphasis on measurements of low-loss materials.

NIST MEASUREMENT SYSTEMS

Broadband Transmission Line Method

Because of its popularity throughout industry and elsewhere, much effort has been devoted to improving the broadband transmission-line technique at NIST.[3,4] In this technique, a precisely machined sample of the material under test is placed within a length of coaxial air line or waveguide. The 7 mm diameter coaxial air line is generally the preferred method, because it is potentially capable of covering a very wide frequency range (up to 18 GHz). The method cannot reliably determine the dielectric and magnetic loss indices, ϵ_r'' & μ_r'' of low-loss materials due to poor measurement sensitivity, but is satisfactory for determining the real parts, ϵ_r'& μ_r'. NIST has successfully used this technique for characterizing ferrites in the spectral range 1-1000 MHz where they exhibit appreciable magnetic properties as well as significant magnetic and dielectric losses. Figure 1 shows measured data for a commercially available nickel-zinc ferrite. Note that the data cover almost 5 decades in frequency (0.3-10 000 MHz) and include error bars at selected frequencies. These are generally worst for permittivity data due to the unavoidable presence of residual air gaps between sample and the coaxial inner conductor, which can lead to serious inaccuracy particularly for high-permittivity materials ($\epsilon' > 50$). This technique is therefore unsuitable for the characterization of high-permittivity ceramics.

Open-ended Coaxial Probe

The open-ended coaxial probe method has advantages of being broadband, fast and nondestructive (i.e. it does not require a precisely machined sample, only a uniformly flat surface) and appears to hold promise for accurately characterizing thin substrates and circuit boards, composed of isotropic polymers of low-to-medium permittivity, when used in either a one- or two-port configuration. A 3.5 mm diameter probe with associated software is commercially available from a major supplier

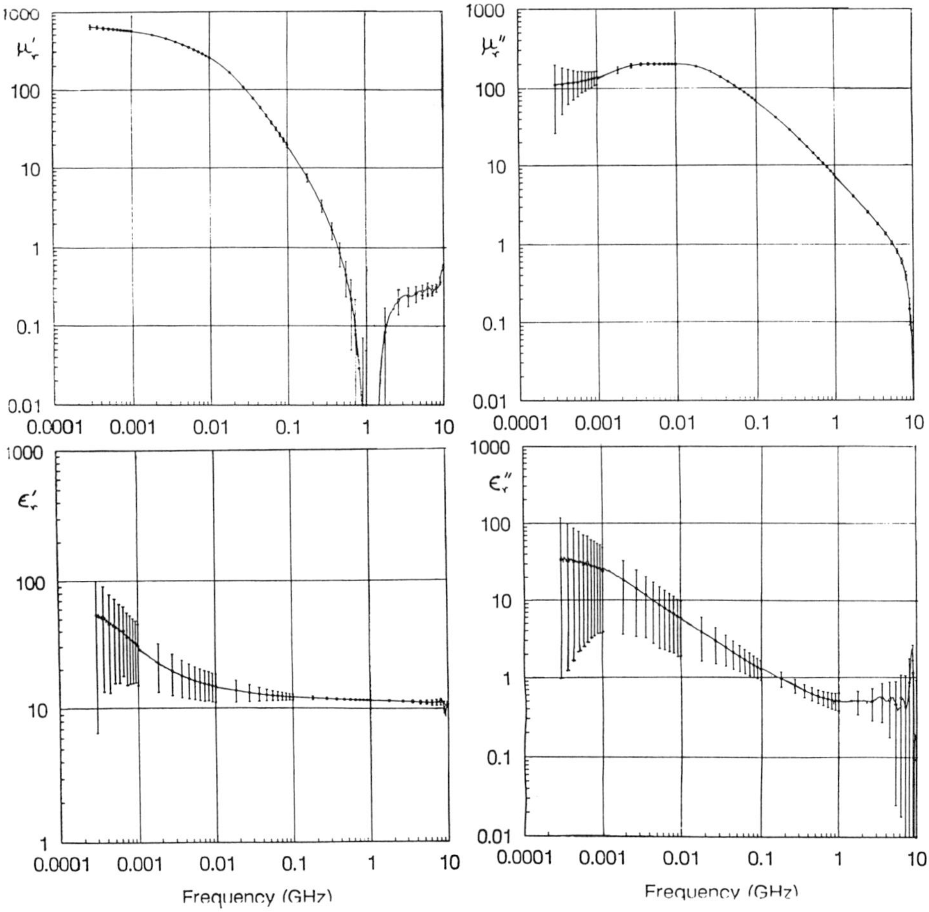

Fig. 1: Complex permeability and permittivity for a Ni-Zn ferrite from 0.3 MHz to 10 GHz

of microwave instrumentation. However, questions remain regarding the probe's accuracy, particularly when used with solid materials of medium-to-high permittivity, due to lack of correction for probe lift-off (the unavoidable presence of an air gap between probe and material surface). Also, available software is based on low-frequency lumped-circuit theory and assumes that the probe is terminated into an infinite half-space of the material under test. Together with many others, NIST has developed advanced full-wave solutions that incorporate both TEM and higher-order TM_{0n} modes. In addition, the

NIST solution includes and corrects for probe lift-off, and incorporates a layered sample of finite thickness, terminated by either a conductive backing (short circuit) or half-space of known material[5,6]. Simultaneous measurements of both complex permeability and permittivity are possible when using the two-port configuration; the one-port configuration requires two separate measurements, involving different layer thickness. The NIST study considered the uncertainty in measured ϵ' as a function of lift-off and frequency and also examined the optimal probe diameter when characterizing materials of different permittivities. Figure 2 shows how the measured permittivity for glass of nominal $\epsilon_r' = 6.15$ varies with lift-off, when measured with a 35 mm diameter probe at 1 GHz. Note the

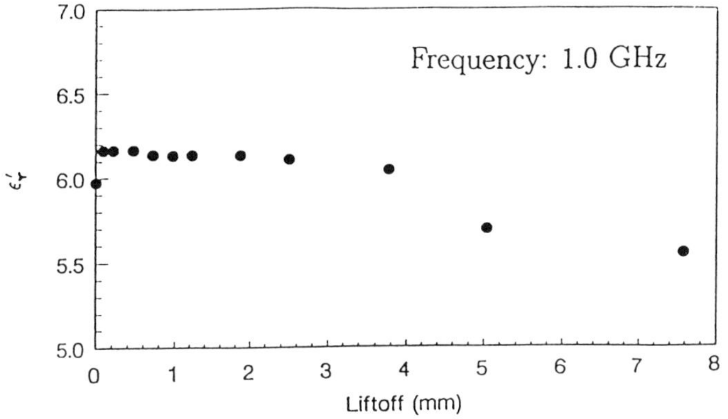

Figure 2: Variation of measured permittivity data with lift-off for glass of nominal $\epsilon_r' = 6.15$

significant drop in measured ϵ_r' value for zero lift-off, representing the uncorrected air-gap error. Like the broadband transmission line methods, the open-ended probe method works best with medium-to-high loss materials. Measurement accuracies can be improved by selecting the optimal aperture diameter for the material properties to be measured and for the desired frequency range.

RF Impedance/Material Analyzer

This is a commercially available technique that is also able to characterize thin (< 3 mm) PC board and substrate materials, as well as ferrites, over the range 1-1800 MHz[7]. The instrument derives the complex permittivity of the substrate material by measuring the capacitance and admittance of a parallel plate fixture, both when empty and when filled with the material under test. Similarly, it derives the permeability of magnetic ferrites by measuring the inductance change when an inductor fixture is filled with a toroidal sample of the material under test. A capability for remotely locating these fixtures allows for measurements over the temperature range of -55 to +200°C. NIST investigated the accuracy and repeatability of this instrument when measuring thin substrates and has concluded that its accuracy is within the range quoted by the manufacturer (10-15% for ϵ_r' in the 10-200 MHz range with some deterioration outside this range). Figure 3 shows a NIST measurement of a typical low-permittivity substrate polymer; the increase in ϵ' that is manifested at frequencies above 800 MHz is an artifact and indicative of increased inaccuracy at upper frequencies. The repeatability measurements performed on PTFE sheets over a period of about ten days showed variability of ±1% about the mean value of $\epsilon_r' = 2.2$

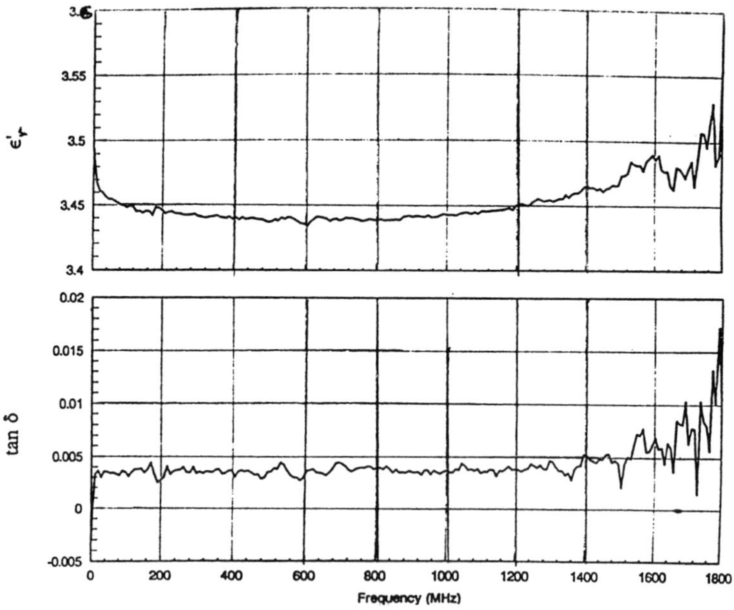

Figure 3: Measurement of a typical polymer substrate material using the RF Impedance/Material Analyzer

NIST 60-mm Diameter TE_{01n} Cylindrical Cavity

Loss-factor measurements for very low-loss materials ($\tan\delta = \epsilon_r''/\epsilon_r' < 10^{-4}$) can only be performed in high-Q resonator systems. Four different resonantor techniques will be discussed here, all of which are in use at NIST. Material characterization using resonator methods basically involves measuring the change in cavity modal parameters of the empty resonator versus that when a material sample is present; ϵ_r' is derived from the measured shift in resonant frequency and $\tan\delta$ is derived from the change in Q-factor (the higher the Q of the empty resonator, the better the sensitivity for measuring low loss-factors.) A major disadvantage of resonators is that they are not broadband; i.e. the measurement is a single frequency one. However, multiple frequency measurements can be performed by retuning the cavity, where possible, or by use of higher-order and harmonic resonances. The accuracy of such measurements depends primarily on how good a field solution is available for the cavity. Mode-matched solutions are the most accurate (typically ±0.2% in ϵ_r', ±5% in ϵ_r''), but cavity perturbation solutions also give very accurate results provided that the limits of the perturbation theory are not exceeded. The reason why the quoted ϵ_r'' accuracy is so much poorer is due to the uncertainty in defining the metal wall losses of the cavity.

The NIST 60-mm diameter cylindrical cavity is a tunable, 2-port resonator, capable of very precise characterization of low-loss, nonmagnetic materials at X-band (8-13 GHz).[8] The walls consist of a helical waveguide which cannot support axial currents. Hence most cylindrical modes, with the exception of the low-loss TE_{0n} family, are attenuated by at least 20 dB. The dominant TE_{01p} mode contains only azimuthally constant electric (E)-fields with no radial component, thereby eliminating any air-gap problems. Because the E-field for this mode is parallel to the end plate, the NIST cavity is restricted to characterizing materials in this plane only. Approximately 17 modes, where p = 15, ... ,31, are available for measurements. The resonator length is adjustable from 408.5 to 433.5 mm. Currently, the cavity is excited through two small waveguide-fed slots in the upper end plate, giving a Q-factor of about 80 000 at 10 GHz. However, the existing slot excitation will soon be replaced by loop excitation which allows for readily altering the degree of cavity excitation and also provides greater operating bandwidth. The sample under test needs to be precisely machined to the resonator's diameter and then placed on the moveable end plate of the cavity. Figure 4 shows some data for a sample of alumina

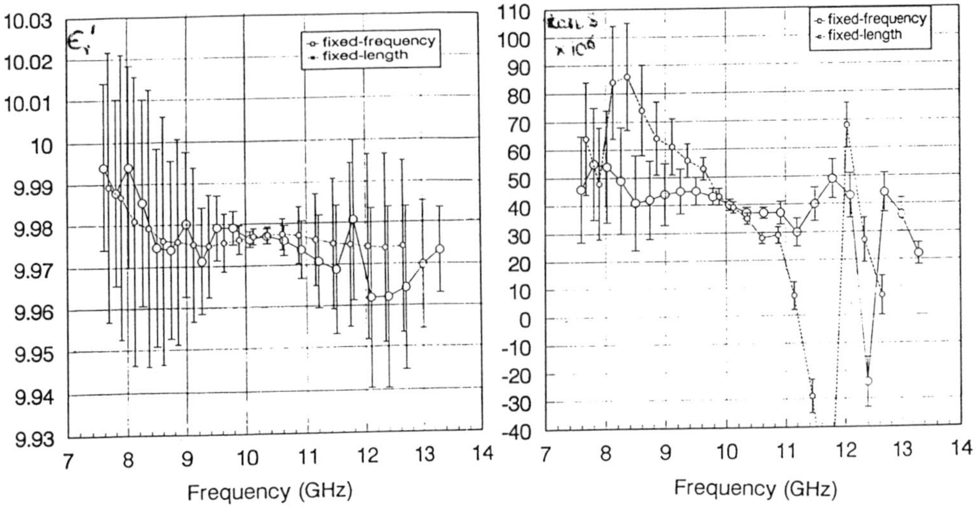

Fig. 4. Relative permittivity and loss factor for alumina, as measured in the NIST cylindrical cavity.

(Al_2O_3), measured in the NIST cylindrical cavity, using both the fixed frequency technique in which the cavity is retuned after material placement and the fixed length (no retuning) technique; the former is considered to be more accurate. Note that loss factors down to 10^{-5} were reliably determined using this fixture.

Cavity Dielectrometer

This is a commercially-available 2-port cylindrical cavity, designed for nondestructive characterization of substrates and is available in diameters of 20.3, 25.4 and 38.1 mm[9]. The cylinder is split into two equal lengths and each half-cylinder is separately excited by loop coupling. The substrate under test is mounted between the two sections and needs to be flat, uniformally thick, unclad (no metallization) and must cover an area somewhat greater than the cross-sectional area of the cavity. Measurements are performed using menu driven software that is supplied with the cavity.

This cavity also employs the low-loss TE_{01p} mode, which greatly reduces air-gap errors. As with the NIST cylindrical cavity, this means that the E-field is parallel to the plane of the substrate and not normal to it. This is only likely to be a problem when measuring anisotropic single-crystal materials; most substrates are composed of polycrystalline ceramics and polymers that are isotropic (it should be possible to perform measurements in this cavity using other modes, in which the E-field is normal to the substrate). Since the cavity is not tunable, the measurement frequency cannot be preset and depends on the permittivity and thickness of the material under test. This frequency generally lies in the range of approximately 2-20 GHz. Corrections are provided in the software to account for error introduced by the sample thickness. Measurements performed by NIST on dielectric reference material samples show that its accuracy is well within the ±1% uncertainty for ϵ_r' claimed by the manufacturer. NIST is currently developing a dielectrically-loaded version of this cavity, which effectively reduces the operating frequency of the system (< 1 GHz).

Dielectric Rod Resonator

In this technique, a cylindrical sample (with radius, a; see Figure 5) of the material under test is placed between two parallel-plate conductors whose diameter is determined, in practice, by the requirement that the evanescent fields outside the cylindrical sample be reduced to a given level[10].

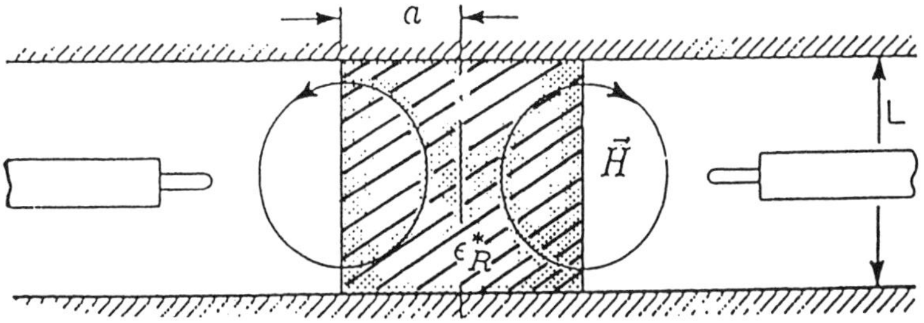

Figure 5: Dielectric Rod Resonator

As with the cavity dielectrometer (see above), the resonant system is usually excited by adjustable coupling loops, giving maximum frequency coverage. Depending on the desired field orientation, measurements can be performed using different resonant modes, including some hybrid ones. For isotropic materials, the preferred mode is again the TE_{0np}. When operated in this mode, the presence of an air gap between sample and conductor plates can be used to substantially alter the resonant frequency but has a negligible effect on measurement accuracy. Mode-matched solutions have been developed for the dielectric rod resonator problem, leading to excellent measurement accuracy that is comparable to other resonator methods (typically ±0.3% for $\epsilon' < 35$). The accuracy for loss factor determination is again dependent on how well the conductor losses of the plates are known; the use of very low-loss superconductors has led to very accurate loss factor measurements in ceramics at cryogenic temperatures[11]. Table 1 shows some typical data obtained for different materials using the dielectric rod technique (the last three materials listed represent cryogenic measurements at 77°K).

Table 1. Measurement Results for the Parallel-Plate Resonator

Material	Frequency (GHz)	Permittivity	Loss Tangent
Fused Silica	9.086	3.824	9.1×10^{-5}
Alumina	3.460	10.04	6×10^{-5}
Rutile	8.825	96.4	4.7×10^{-4}
PTFE	10.267	2.07	1.6×10^{-4}
PTFE	15.278	2.06	1.8×10^{-4}
PTFE	20.457	2.06	1.8×10^{-4}
Lanthanum Aluminate (Crystalline Form)	17.216	23.909	5.95×10^{-5}
Magnesium Oxide (Crystalline Form)	15.079	9.782	2.8×10^{-5}
Neodymium Gallate (Crystalline Form)	17.184	21.822	1.6×10^{-4}

Re-entrant Cavity Method

This is basically a tunable coaxial cavity in which the sample under test is placed in a gap between two sections of the center conductor[12]. The technique, which was first developed many decades ago, is of importance because it promises to provide accurate characterization of substrates in the mid-frequency range, 100-1000 MHz where most other resonator techniques cannot be used. For applications involving the characterization of anisotropic materials, the technique has the additional advantage of providing a field orientation that is normal to the plane of the substrate. NIST is currently working on improved full-wave field solutions for this device, with corrections for air-gap effects, which promise to provide accuracies of the order of ±2%.

REFERENCE MATERIALS

Considerable effort has been devoted to identifying and thoroughly characterizing suitable dielectric and magnetic reference materials at NIST. In general, these need to be solid materials that are linear, spatially homogeneous, and isotropic, as well as minimally dispersive[13]. They also need to be very temperature stable in their physical and electromagnetic properties, chemically inert, and nonhygroscopic. It is also desirable that these materials be readily available, of modest cost, and easily machined.

Low-loss dielectric materials used as NIST dielectric reference materials are tabulated below, together with relative permittivity and loss factor data at 10 GHz:

TABLE 2: NIST DIELECTRIC REFERENCE MATERIALS*

	ϵ_r'	$\tan\delta$
Cross-linked Polystyrene (Rexolite)	2.53	0.0024
Fused Quartz (Corning 7940)	3.82	0.0003
Soda-Lime Glass (NIST SRM 710a)	6.82	0.049
Alumina (Coors Vistal)	9.97	0.00004
Mg-Ti Ceramic (Trans-Tech D-16)	16.05	0.0014

Magnetic reference materials pose many more problems than do their dielectric counterparts primarily because they exhibit much more batch-to-batch and sample-to-sample variability; it is also important to ensure that they are demagnetized (free of static bias). These materials exhibit strong magnetic properties at RF frequencies

* Trade names are used in order to adequately specify the material for the reader. This does not imply any endorsement by NIST; other materials are equally satisfactory.

below 1 GHz, but are only weakly diamagnetic above this; magnetic properties are very dispersive in both frequency and temperature. Several commercially made ferrites and loaded polymers have been provisionally selected for use as NIST magnetic reference materials.

NIST MEASUREMENT SERVICES, INTERCOMPARISONS, AND DATABASE

NIST is now routinely providing "special test"measurement services to industrial customers and others for the microwave characterization of customer furnished materials. Machining of the sample is performed either by NIST or the customer depending on the customer's preference. NIST also supplies machined and unmachined samples of dielectric reference materials, as needed.

The NIST materials project has also recently completed a national intercomparison on low-loss nonmagnetic materials measurements using the 7-mm diameter coaxial air line technique. Four samples of NIST dielectric reference materials were circulated among ten participating industrial laboratories. Data from this intercomparison were recently published[14]. Our preliminary conclusion is that interlaboratory agreement ranged from fair to poor, depending on material permittivity. A similar national intercomparison involving magnetic materials is currently underway.

To date, little effort has been devoted to developing a NIST database for material parameters. This will probably be done later, when more baseline data have been accumulated on some of the basic materials that are being measured at NIST.

CONCLUSIONS

Significant advances have been realized in the NIST EPM Program. Important improvements have been made in transmission line, open-ended probe, and cylindrical cavity resonator measurements. Work on resonator techniques that can be used for characterizing thin films and substrates in the mid-spectrum region are currently being emphasized. Future work will involve both cryogenic and elevated temperature measurements as well as millimeter wave free-field resonator measurements.

ACKNOWLEDGEMENTS

The author gratefully acknowledges the contributions of all current and past members of the NIST EPM Project: J.R. Baker-Jarvis, R.G. Geyer, J.H. Grosvenor, Jr., M.D. Janezic, C.A. Jones, Y. Kantor, W.A. Kissick, J. Krupka, B.F. Riddle and E.J. Vanzura. Thanks are also expressed to R.B. Marks and W.A. Kissick for manuscript reviews.

REFERENCES

1. C.M. Weil, "Electromagnetic Properties of Materials: the NIST Metrology Program," Proceedings of the Fifth DOD Electromagnetic Window Symposium, Vol 1, Boulder CO, October 1993. Published by Materials Directorate, Wright Laboratory, USAF Materiel Command, Wright Patterson AFB, OH 45443-7734; Report No. WL-TR-93-4117

2. C.M. Weil in Microwave Processing of Materials III, ed. R.L. Beatty, W.H. Sutton, M.F. Iskander (Mat. Res. Soc. Symp. Proc, 269), April 1992, p. 517.

3. J. Baker-Jarvis, "Dielectric and Magnetic Measurement Methods in Transmission Lines: An Overview," Proceedings 1992 AMTA Workshop, p. 1-1; published by Antenna Measurements Techniques Assoc.

4. J. Baker-Jarvis, M.D. Janezic, J.H. Grosvenor & R.G. Geyer, "Transmission/Reflection and Short-Circuit Line Methods for Measuring Permittivity and Permeability," NIST Technical Note 1355-R, Boulder, CO, December 1993.

5. J. Baker-Jarvis, M.D. Janezic, P.D. Domich & R.G. Geyer, "Analysis of an Open-ended Coaxial Probe with Lift-Off for Nondestructive Testing," IEEE Trans. Instrumentation & Meas., IM-43 (5), pp 711-718, Oct. 1994

6. J. Baker-Jarvis & M.D. Janezic, "Open-ended Coaxial Probes for Nondestructive Testing of Substrates and Circuit Boards," In Microwave Processing of Materials IV, Mat. Res. Soc. Symp. Proc., 347, pp 215-220, April 1994.

7. HP Solution Note 4291-4, "Permittivity Measurements of PC Board and Substrate Materials using the HP 4291A and HP-16453A" and HP Solution Note 4291-5, "Permeability Measurements using the HP 4291A and HP 16454A." Published by Hewlett-Packard Company, 1994.

8. E.J. Vanzura, R.G. Geyer & M.D. Janezic: "The NIST 60-mm Diameter Cylindrical Cavity Resonator: Performance Evaluation for Permittivity Measurements," NIST Technical Note 1354, Boulder, CO, August 1993.

9. G. Kent: "A Dielectrometer for the Measurement of Substrate Permittivity," Microwave Journal, 34 (12), pp 72-82, December 1991.

10. Y. Kobayashi & M. Katoh: "Microwave Measurement of Dielectric Properties of Low-Loss Materials by the Dielectric Rod Resonator Method," IEEE Trans Microwave Theory & Techs, MTT-33 (7), pp 586-592 (July 1985)

11. J. Krupka, R. Geyer, M. Kuhn & J.H. Hinken: "Dielectric Properties of Single Crystals of Al_2O_3, $LaAlO_3$, $NdGaO_3$, $SrTiO_3$ and MgO at Cryogenic Temperatures," Ibid, MTT-42 (10), pp 1886-1890 (October 1994).

12. C.N. Works, T.W. Dakin & F.W. Boggs: "A Resonant Cavity Method for Measuring Dielectric Properties at Ultrahigh Frequencies," AIEE Trans, 63, pp 1092-1098 (1944)

13. R.G. Geyer, "Dielectric Characterization and Reference Materials," NIST Technical Note 1338, Boulder, CO, April 1990.

14. E.J. Vanzura, J. Baker-Jarvis, J. Grosvenor & M. Janezic: "Intercomparison of Permittivity Measurements using the Transmission/Reflection Method in 7 mm Coaxial Transmission Lines," IEEE Trans Microwave Theory & Techs, MTT-42 (11), pp 2063-2070 (November 1994).

ACCURATE MEASUREMENTS OF ELECTRICAL PROPERTIES OF DIELECTRIC RESONATORS AND SUBSTRATE MATERIALS

Kawthar A. Zaki and Chi Wang

Electrical Engineering Department
University of Maryland
College Park, MD 20742

ABSTRACT

A nondestructive method for accurate characterization of dielectric resonators (D.R.'s) and dielectric substrate materials is described. The method uses cylindrical or rectangular metallic enclosure to hold the D.R. or the substrate sample for measurements. Extraction of relative permittivity (ϵ_r) and loss tangent ($tan\delta$) is performed using an accurate powerful electromagnetic mode matching simulation model, which can easily be implemented on the control computer of an automatic network analyzer. Thus the whole process can be automated. The analysis program allows the extraction of properties for isotropic as well as anisotropic materials.

Simulation results of several examples of different types and shapes, of D.R.'s (solid resonators with holders, ring type resonators with holders, and both types of D.R.'s on substrate) are presented.

I. INTRODUCTION

The objective of the work described in this paper is to develop an accurate method (accuracy better than 0.5%) for the precise characterization of the electrical properties of dielectric resonators and substrate materials, of different shapes and sizes, isotropic or anisotropic. The basic principal of the method is well known, but its practical implementation, its high precision, its ability to characterize anisotropic materials and its speed of processing are the main features that distinguish it from previous work.

Measurements of ϵ_r, but not $tan\delta$ of bulk cylindrical dielectric resonators was described by Cohn etal [1]. Accuracy of the measurements was estimated to be 5% to 10%. For substrate characterization, Napoli etal [2] used a metalized rectangular substrate which formed a rectangular waveguide with open side walls. Radiation from the edges of the substrate limited the accuracy of this method to about 10%, and the loss tangent of the material could not be measured. To improve the accuracy, J.Q. Howell [3] used conductive paint on the edges of the metalized substrate to eliminate radiation. Methods based on the use of microstrip cavity was described by Itoh [4] and A.R. Gerhard [5]. The microstrip resonator was placed on the substrate. The effects of the contact between the D.R. and substrate, the effects of the microstrip line, as well as the effects of the gap between the line and resonator all produce some measurement errors, which are unknown and not easily quantifiable. The use of a TM_{010} mode cavity with the resonator perturbing its fields was used for the measurement of the dielectric constant [6]. Accuracy of the perturbation method is estimated to be about 5%. For substrate measurements various transmission line techniques were used [7], [8]. Methods based on the measurements of several resonant frequencies of higher order modes of a cylindrical resonator sandwiched between parallel plates were described by Wheless etal [9] and by Kobayashi etal [10]. These methods used a minimization of an error function constructed as the root mean square of the sum of differences between the measured resonances of several higher order modes and those computed. Again these methods have sources of inaccuracy in the contacts with the resonator, and the possibility that ϵ_r may vary with frequency. An evanescent mode rectangular cavity was used in [11], a cylindrical cavity in [12] to measure ϵ_r of resonators. Finally a coaxial dielectric discontinuity was used to measure ϵ_r of a ring resonator [13]. Accuracy limitations of the current measurement techniques are due principally to two factors: first the mechanical configuration uncertainity (i.e. effects of dielectric to metallic interface, or dielectric to dielectric interfaces) and second the precision by which the measured configuration can be modeled and simulated.

The work described in this paper alleviates both of these limitations. It is based on a highly flexible and fast electromagnetic simulation coupled with precision frequency measurements to achieve high accuracy of dielectric material properties. The methods of measurement and the simulation are designed to accommodate a large number of configurations, described in Section II, and the samples are not altered to perform the measurements.

In Section III, the methods of simulation used to extract the material properties are discussed. The methods are based on mode matching and both rectangular and circular cylindrical boundaries are treated. The simulations include the effects of anisotropy, losses in the enclosures and losses in the dielectric materials. Field distributions of the resonant modes in the structures are presented and are used to show how the measurements fixtures are designed to reduce or eliminate measurements uncertainities associated with interfaces.

The measurement procedures are described in Section IV. The method of calibration of the fixtures and measurements of the resonant frequency and unloaded Q's are discussed. The procedures of extraction of the dielectric material properties ϵ_r and $tan\delta$ from the measured data using the simulation programs are described. Several examples and actual test results are presented.

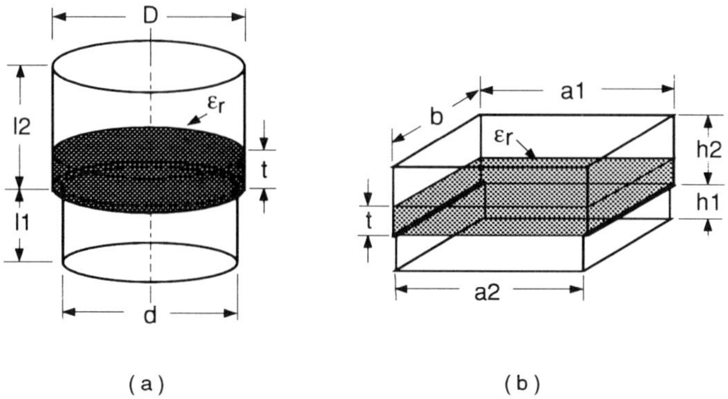

Fig. 1 Two configurations of the dielectric resonator or substrate material in a closed cavity (a) cylindrical cavity (b) rectangular cavity.

II. MEASUREMENT CONFIGURATIONS:

For accurate measurements of both ϵ_r and $tan\delta$ of resonators and substrate materials, closed cavity structures with the samples loaded inside the cavity are the best choice. Figure 1 shows two configurations: (a) a cylindrical cavity and (b) a rectangular cavity.

The ϵ_r measurement is achieved by first measuring the resonant frequency of the cavity with the material inside it, in a specific mode. The value of ϵ_r is then extracted from the measured resonant frequency by simulating the electromagnetic fields in the cavity for the specific mode being excited, and varying ϵ_r in the simulation until the resulting resonant frequency is equal to the measured resonant frequency. For accurate measurements, a small change in ϵ_r should produce a large change in the resonant frequency (i.e. $\Delta\epsilon/\Delta f$ must be small). This is achieved by judicious choice of the cavity dimensions and the resonant mode employed for measurement.

The $tan\delta$ measurement is achieved by measuring the unloaded Q of the resonator in the same mode used for ϵ_r measurement. This unloaded Q is due to the combined loss of the dielectric material loss (P_d) and the metallic enclosure loss (P_e). The dielectric material tanδ can be computed from the measured unloaded Q, and a knowledge of the surface resistivity of the metallic enclosure. For accurate measurement of tanδ, the contribution of the dielectric loss P_d to the total Q must be significant, i.e. $\Delta Q/\Delta P_d$ (or P_d/P_e) must be large. This implies that the metallic enclosure walls should be far away from the sample to minimize P_e, and a good conductor with low surface resistivity (i.e. copper or silver plated metal) should be used.

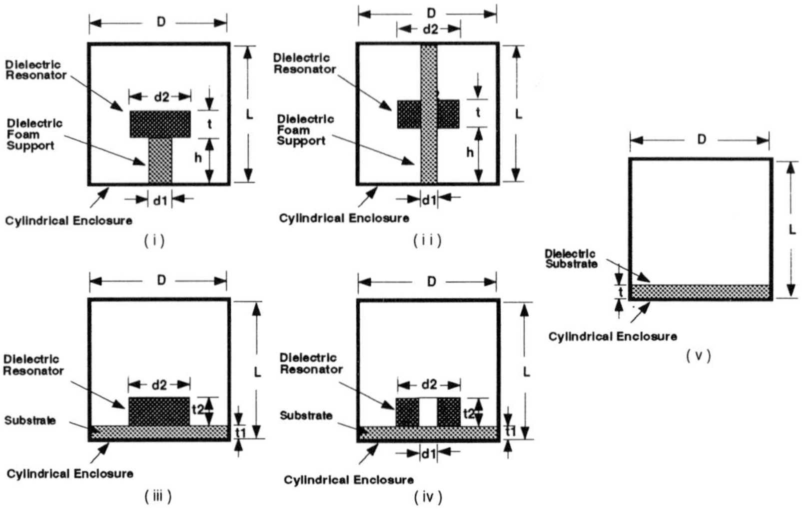

Fig. 2 Examples of cylindrical structures that can be analyzed by the general simulation program

Comparing the requirements of high accuracy for ϵ_r measurements and those for tan δ measurements, it is found that making $\Delta\epsilon/\Delta f$ small will automatically reduce $\Delta Q/\Delta P_d$, and vise versa. Therefore, a trade off has to be made if the same structure (holder) is to be used for both ϵ_r and tanδ measurements. In most cases, however, ϵ_r measurements can be made with higher accuracy, than tan δ. Thus it is important to optimize the structure for accurate tanδ measurements, and the resultant ϵ_r accuracy is usually acceptable.

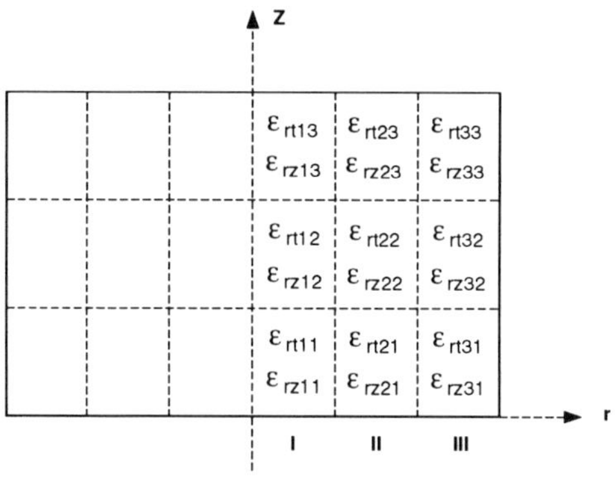

Fig. 3 Configuration of the three layer dielectric resonator in a cylindrical cavity

The shapes that can be used for the measurements of the material properties must satisfy the condition that the geometrical configuration can be analyzed by mode matching, so that sufficient accuracy can be obtained. There is a large class of cylindrical structures for which the mode matching analysis has been implemented in a general simulation program and thus all these configurations may be used for measurements. The configurations are shown in Fig. 2 and include the following cases:

(i) Solid dielectric resonator with support
(ii) Ring dielectric resonator with support
(iii) Solid dielectric resonator on substrate
(iv) Ring dielectric resonator on substrate
(v) Circular substrate alone.

III. MODELING OF FIELDS IN MEASUREMENT STRUCTURE

Resonant Frequencies and Field Distribution

Mode matching technique is a widely used method for modeling microwave transmission and resonant structures. Mode matching provides all necessary information about the field distribution of the modes that can exist in the structures accurately and efficiently. General computer programs have been developed to model the structure shown in Fig. 3. The perfectly conducting enclosure is filled with three layers of dielectric materials, each layer has three radial regions, for a total of nine annular regions. The material in each region is assumed to be anisotropic with two different dielectric constants: one in the axial direction ϵ_{ijz} and another in the transverse direction ϵ_{ijt}, where $i, j = 1, 2, 3$. Details of the mode matching method for the analysis of this structure can be found in Reference [14] and is summarized in the following steps. The solution assumes either no azimuthal field variation (TE or TM-modes) or the same ϕ variation (sin$n\phi$ and cos$n\phi$) for all the fields (hybrid modes):

(i) Find the eigen modes for each of the planar parallel plate waveguides with three dielectric layers. These eigen modes are solutions of Maxwell's equation that satisfy the boundary conditions at the interfaces between the dielectric layers, the top and bottom conducting planes.

(ii) Express the solution in each of the three radial regions as summations of products of the parallel plate solutions obtained in step (i) above, and appropriate cylindrical functions.

(iii) Match the radial boundary conditions at the interfaces between the regions in the radial direction, by equating the tangential electric and magnetic fields at the boundaries. Match the tangential field to be zero at the outside cylindrical surface.

(iv) Take inner products of the resulting boundary conditions in (iii) with the eigen modes at the interfaces. This results in an infinite system of homogeneous linear equations with the field expansion coefficients as the unknowns.

(v) Truncate the infinite homogeneous system in (iv) to a finite number of equations. The resonant frequency of the structure is obtained by solving the equation resulting from equating the determinant of the matrix of the system of equations to zero.

(vi) once the resonant frequency is obtained it is substituted back into the linear truncated system of equations, which can be solved for all the field expansion coefficients in terms of a single coefficient.

(vii) With the field expansion coefficients known all the fields in the structure can be obtained.

A similar procedure is used to analyze rectangular resonant structures.

Q calculations

Determination of the unloaded Q involves the calculation of the energy (W) stored in the structure and the power (P) lost in the metallic enclosure and the dielectric materials. The unloaded Q is calculated from:

$$Q = \frac{\omega_0 W}{P} = \frac{1}{\frac{1}{Q_d} + \frac{1}{Q_e}} \tag{1}$$

where ω_0 is the resonant frequency, Q_d and Q_e are the unloaded Q's corresponding to the losses in the dielectric material and the enclosure surfaces, respectively. The stored energy includes both the energy stored in the electric fields (W_e) and the magnetic fields (W_m) at resonance: i.e.

$$W = W_e + W_m = 2W_e = 2W_m \tag{2}$$

where

$$W_e = \frac{1}{4}\int_V \vec{D} \cdot \vec{E}^* dV \tag{3}$$

$$W_m = \frac{\mu}{4}\int_V \vec{H} \cdot \vec{H}^* dV \tag{4}$$

where

$$\vec{D} = \epsilon \vec{E} \tag{5}$$

and ϵ is the permittivity of the anisotropic material.

The power loss is obtained from:

$$P = P_e + P_d \tag{6}$$

where P_e is the power loss in the enclosure walls, and P_d is the power loss in the dielectric material. The power losses are computed based on the low-loss assumptions that the fields are unperturbed by the losses. The surface current density flowing on enclosure walls is $\hat{n} \times \bar{H}$, where $\hat{n}$ is the unit normal to the enclosure surface, and $\bar{H}$ is the magnetic filed. The power loss due to the enclosure surface current is thus:

$$P_e = \frac{R_s}{2}\oint_s (\hat{n} \times \vec{H}) \cdot (\hat{n} \times \vec{H}^*) ds \tag{7}$$

where

$$R_s = \sqrt{\frac{\omega_0 \mu_0}{2\sigma}} \tag{8}$$

is the surface resistivity of the metallic enclosure, and σ is the surface conductivity of the metallic enclosure.

The power lost in the dielectric material is obtained from the electric energy stored in the dielectric medium as:

$$P_d = \sum_i \frac{\sigma_i'}{2}\int_{V_i} \vec{D} \cdot \vec{E}^* dV \tag{9}$$

where σ_i' is the conductivity of the material in medium i , given by:

$$\sigma_i' = \omega_0 tan\delta_i \tag{10}$$

where tan δ_i is the loss tangent of the material in medium i.

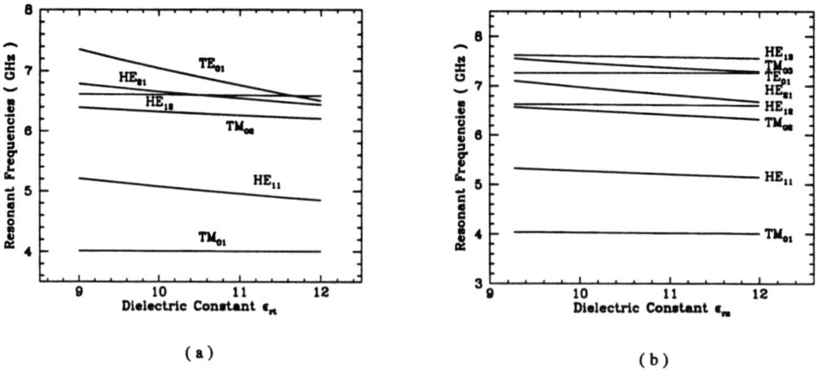

Fig. 4 The effect of varying the radial and axial relative dielectric constant of an anisotropic resonator on the resonant frequencies of several modes. D=2.00", L=1.00", d2=0.760", t=0.230"

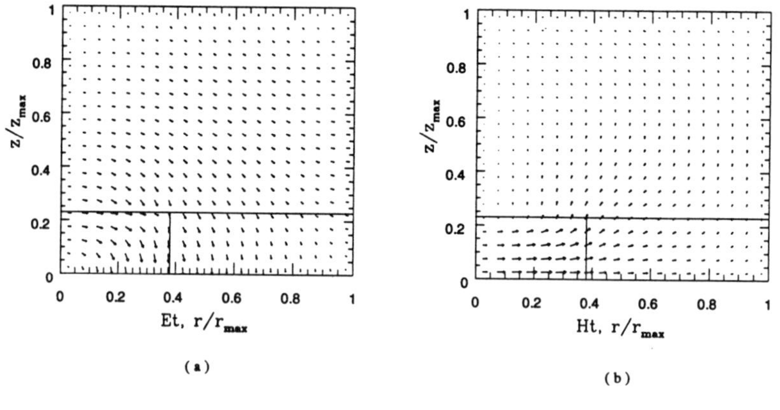

Fig. 5 The field distributions of the HE_{21} mode in $z - r$ plane.

A computer program was developed to implement the mode matching analysis of the anisotropic resonators (shown in Fig. 3) and to compute the unloaded Q's of the structure. The program is used for the extraction of the relative dielectric constant of the material from the measurements of the resonant frequency. The program was used extensively to simulate the effects of varying the structure parameters on the resonant frequencies and field distribution of various modes. It was also used to estimate the effect of the anisotropy of the dielectric material on the resonant frequencies of the structure. Some sample data obtained from this program are presented below.

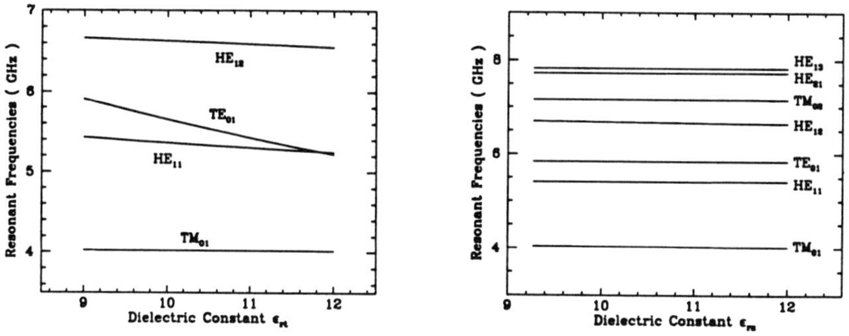

(a) (b)

Fig. 6 The effect of varying the radial and axial relative dielectric constant of anisotropic resonator with support of $\epsilon_r = 4.5$ on the resonant frequencies of several modes. D=2.00", L=1.00", d2=0.760", t=0.230", d1=0.40", h=0.385"

Fig. 4 shows the effect of varying the radial relative dielectric constant ϵ_t ($\epsilon_z = 11.5$) and the axial relative dielectric constant ϵ_z ($\epsilon_t = 9.27$) of anisotropic resonator on the resonant frequencies of several modes. In this figure the resonator is placed directly on the bottom of the enclosure with a foam of dielectric constant $\epsilon_r = 1.10$ around the side of the dielectric resonator. It is seen that the resonant frequency varies significantly as ϵ_z is changed. This is particularly significant for the modes which have strong E_z component, such as the HE_{21} mode, whose field distributions is shown in Fig. 5. The effect of the anisotropy is greatly reduced if the D.R. is placed on a support of low ϵ_r. The resonant frequency for a resonator with support as a function of ϵ_t and ϵ_z is shown in Fig. 6. The field distribution of the HE_{11} mode with the resonator support is shown in Fig, 7. The field in the resonator is entirely transverse to the z -direction as can be seen in Fig. 7-a. Thus the variation of ϵ_z does not affect the resonant frequency.

For substrate material measurements the cavities of Fig. 8-a and Fig. 8-b were constructed to accommodate circular and rectangular substrate samples, respectively. Each of these cavities have small "step" to support the substrate. The cavity size is designed to have either TE-mode or TM-mode resonance near the frequency range of interest. The TM -modes have fields in the dielectric substrate perpendicular to the substrate large dimensions, similar to microstrip lines (as shown in Fig. 9). The TE-modes have fields components in transverse plane, similar to coplanar waveguides, as shown in

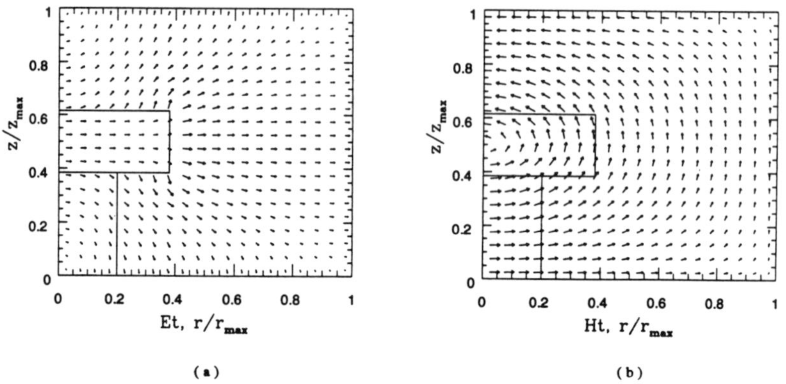

Fig. 7 The field distributions of the HE_{11} mode in $z - r$ plane.

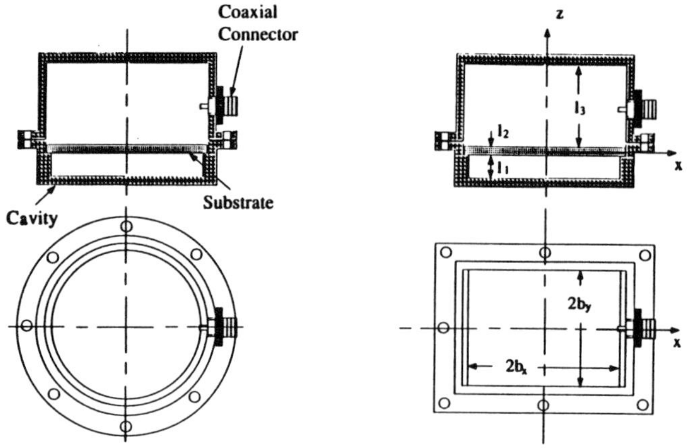

Fig. 8 The structures of the cavities for measuring the substrate samples

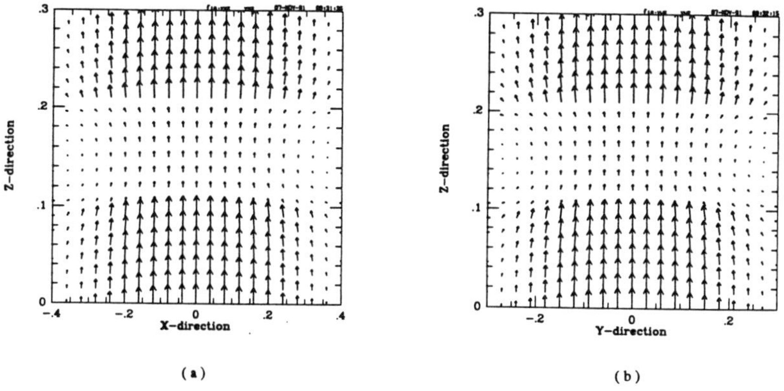

Fig. 9 The field distributions of the TM mode

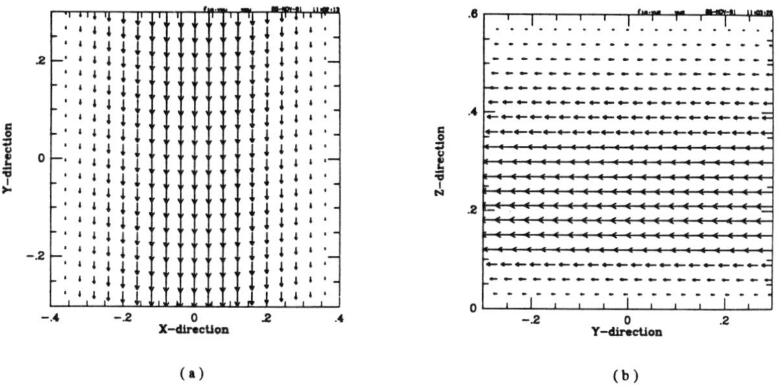

Fig. 10 The field distributions of the TE mode

Fig. 10. The unloaded Q-measurements for the case of the TM-modes was not very accurate, because the cavity was cut in a direction perpendicular to the current flow, and hence caused a large loss that dominated the Q-measurements.

IV. MEASUREMENT PROCEDURES

There are two methods to measure the cavity resonant frequencies and quality factors (Q). The first method (the reflection method) uses measurement of the reflection coefficient (return loss) in a certain frequency range around the resonant frequency of the mode of interest. This method needs only one port measurements, as shown in Fig. 11. The second method (the transmission method) uses measurements of the transmission coefficient (insertion loss) through a two port cavity resonator around the resonant frequency of the mode of interest. This method needs two port transmission measurements as shown in Fig. 12. Detail of the basic measurement techniques for both these methods and their accuracy can be found in Reference [14]. The same techniques are adopted for implementation on an automatic network analyzer

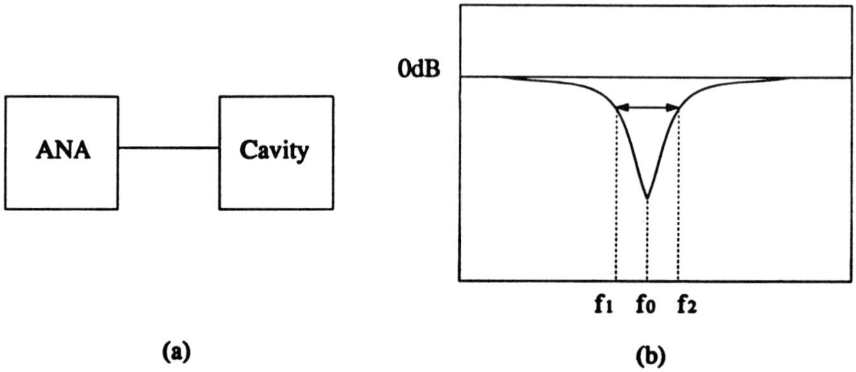

Fig. 11 The step-up and the response for reflection method

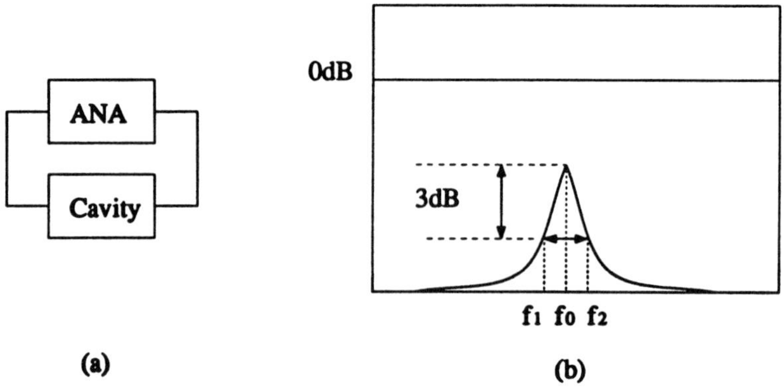

Fig. 12 The step-up and the response for transmission mode

As an example of the measurement procedure the cylindrical cavity of Fig. 8-a was used to measure the properties of several substrate samples. The resonant mode used for the samples characterization was the TE_{111} mode. Typical expected resonant frequencies of the TE_{111} mode and few of the other closest modes are shown for two samples in Fig. 13. To measure the surface conductivity of the enclosure, first the empty holder was used as a cavity and its resonant frequency was measured to be 12.1445 GHz, while the simulated frequency of the same mode was 12.1472 GHz, a difference of only 0.022%. The measured unloaded Q of the empty cavity was 7898. Using the simulation program, the enclosure surface conductivity was computed to be 3.20384 $\times 10^{-4}$ mhos/m. When the sample was inserted in the holder, the measurements of the resonant frequency and unloaded Q were 10.0016 GHz and 1491 respectively. Again using the simulation program the dielectric constant and tanδ of the sample were found to be 3.5948 and 8.904×10^{-4} respectively. Error analysis showed that these values are accurate within $\pm 0.2\%$.

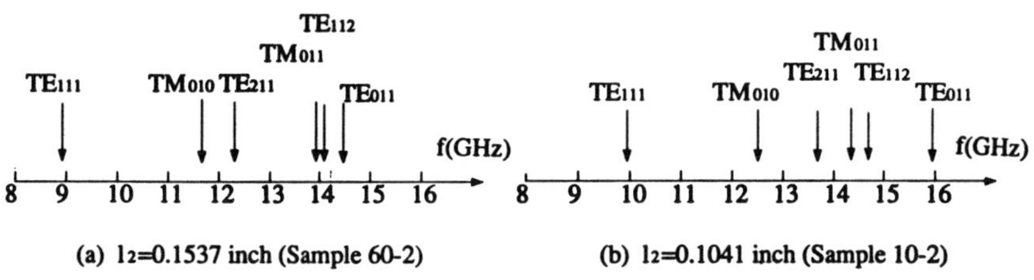

Fig. 13 The expected resonant frequencies

V. CONCLUSION

Measurement procedures are presented for accurate characterization of the electrical properties of dielectric materials. The key to the success of this procedure is the ability to simulate accurately and efficiently the electromagnetic fields in the resonant structure under measurements. This was done with a computer program using the mode matching technique.

The program computes the resonant frequency of the structure and varies the value of the dielectric constant of the material until the simulated frequency is equal to the measured resonant frequency. A complete error analysis of possible sources of errors was investigated, including the sample dimension uncertainity, the enclosure dimensional uncertainity, the frequency measurements and return loss measurement errors and simulation error. Both systematic and random error sources were investigated, and bounds on all the parameters were obtained that provide accuracy of better than 0.5% in the value of ϵ_r and tanδ.

REFERENCES

[1] S. B. Cohn and K. C. Kelly, "Microwave measurement of high-dielectric-constant materials," *IEEE Trans., Microwave Theory and Tech.*, vol. MTT-14, pp. 406-410, September 1966.

[2] L. S. Napoli and J. J. Hughes, "A simple technique for the accurate determination of the microwave dielectric constant for microwave integrated circuit substrates," *IEEE Trans. Microwave Theory and Tech.*, vol. MTT-19, pp. 664-665, July 1971.

[3] J. Q. Howell, "A quick accurate method to measure the dielectric constant of microwave integrated-circuit substrates," *IEEE Trans. Microwave Theory and Tech.*, vol. MTT-21, pp. 142-143, March 1973.

[4] T. Itoh, "A new method for measuring properties of dielectric materials using a microstrip cavity," *IEEE Trans. Microwave Theory and Tech.*, vol. MTT-22, pp. 572-576, May 1974.

[5] A. R. Gerhard, "Measuring dielectric constant of substrates for microstrip applications," *IEEE Trans. Microwave Theory and Tech.*, vol. MTT-24, pp. 485-487, July 1976.

[6] R.M. Pannell and B.W. Jervis, "Two simple methods for the measurement of dielectric permittivity of low-loss microstrip substrate, ", *IEEE Trans. Microwave Theory and Tech.*, vol. MTT-29, pp. 383-386, Apr. 1981.

[7] L. P. Ligthart, "A fast technique for accurate permittivity determination using transmission line methods," *IEEE Trans. Microwave Theory and Tech.*, vol. MTT-31, pp. 249-254, March 1983.

[8] P. A. Bernard and J. M. Gautry, "Measurement of dielectric constant using a microstrip ring resonator," *IEEE Trans. Microwave Theory and Tech.*, vol. MTT-39, pp. 592-595, March 1991.

[9] P. Wheless, Jr. and D. Kajfez, "The use of higher resonant modes in measuring the dielectric constant of dielectric resonators," *1985 IEEE Trans. MTT-S Digest*, pp. 473-476.

[10] Y. Kobayashi and M. Katoh, "Microwave measurement of dielectric properties of low-loss materials by the dielectric rod resonator method," *IEEE Trans. Microwave Theory and Tech.*, vol. MTT-33, pp. 586-592, July 1985.

[11] G. Kent, "An evanescent-mode tester for ceramic dielectric substrates," *IEEE Trans. Microwave Theory and Tech.*, vol. MTT-36, pp. 1451-1454, October 1988.

[12] Y. Kobayashi and T. Senju, "Resonant modes in shielded uniaxial-anisotropic dielectric rod resonators," *IEEE Trans. Microwave Theory and Tech.*, vol. MTT-41, pp. 2198-2205, December 1993.

[13] D. Misra, *et al*, "Noninvasive electrical characterization of materials at microwave frequencies using an open-ended coaxial line: Test of an improved calibration technique," *IEEE Trans. Microwave Theory and Tech.*, vol. MTT-38, pp. 8-14, January 1990.

[14]C.G. Montgomery, "Techniques of Microwave measurements" McGraw HillBook Co., New York, N.Y. 1947.

MICROWAVE TECHNIQUES FOR MEASURING COMPLEX PERMITTIVITY AND PERMEABILITY OF MATERIALS

P. Guillon
I.R.C.O.M. - Faculté des Sciences
123 Av. Albert Thomas
87060 Limoges Cédex - France

INTRODUCTION

Different materials are of fundamental importance to the aerospace, microwave, electronics and communications industries, and include for example microwave absorbing materials, antennas lenses and radomes, substrates for MMIC and microwave components and antennaes.

Basic measurements for the complex permittivity and permeability of those homogeneous solid materials in the microwave spectral region are described including hardware, instrumentation and analysis. Elevated temperature measurements as well as measurements intercomparisons, with a discussion of the strengths and weaknesses of each techniques are also presented.

MEASUREMENTS TECHNIQUES

The major electrical properties of materials that are of interest are resistivity (or conductivity), complex permittivity ($\varepsilon = \varepsilon' - j\,\varepsilon''$) and permeability ($\mu = \mu' - j\,\mu''$). Material losses are usually described in the form of the loss tangent commonly called $\tan\delta_e = \dfrac{\varepsilon''}{\varepsilon'}$. As in the case of permittivity, the losses in a magnetic material are usually expressed in terms of the loss tangent $\tan\delta_m = \dfrac{\mu''}{\mu'}$

There are many ways to measure the electrical parameters of materials and this paper presents the different techniques which could be used and the relative range of permittivity and permeability for which each method is optimum.

Some methods require that the material sample be machined or formed into a specific size and shape, and after the measurement has been made the material sample cannot be used for its original purpose. These methods are called destructive test methods.

Non destructive test methods allow the material to be measured in its raw form, and the sample can be used after it has been measured. These methods are particularly useful in production environments as quality control procedures or as incoming inspection procedures.

The different techniques used to characterize permittivities and permeabilities are classified into two groups, namely :

- resonant techniques
- reflection and transmission techniques

1 - The Resonant Test Methods

The operation is quite simple since the calculation of the permittivity and permeability is deduced from both the measurement and the exact theoretical analysis of the resonant frequency and quality factor of the resonant device.

Very precise measurements can be made with a well constructed cylindrical cavity operating in the $TE_{0,1,2p}$ mode for permeability evaluation and in the $TE_{0,1,2p+1}$ mode for permittivity evaluation. Note that the sample under test is introduced in the cavity (figure 1) in such a way that its diameter is larger than that of the metallic resonator [1].

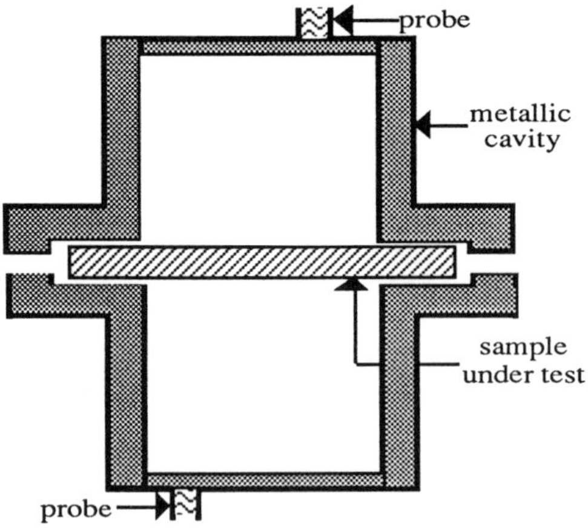

- Figure 1 -

The parallel plate dielectric resonator [2] (figure 2) is convenient for making measurements of high ε' and low loss and anisotropic materials. The mode of that resonator is computed by using the well known mode matching technique, and the TE_{011} mode is the most useful for making measurements.

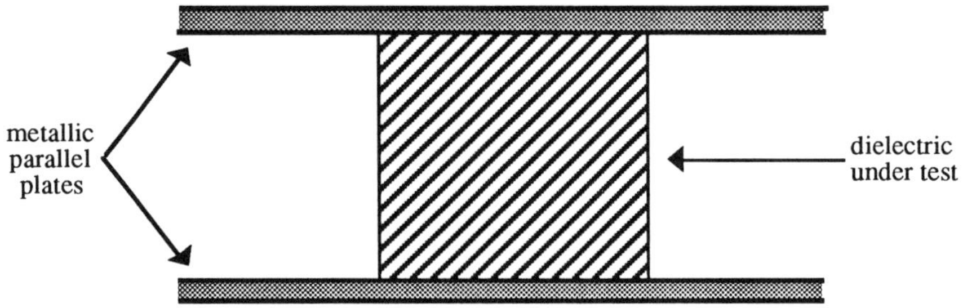

- Figure 2 -

The open cavity dielectric resonator [3] (figure 3) operating in the $TE_{01\delta}$ mode is another method for making precise measurements of the complex permittivity of low loss materials. The resonant frequency and Q factor of that resonator are computed here by using the finite element method.

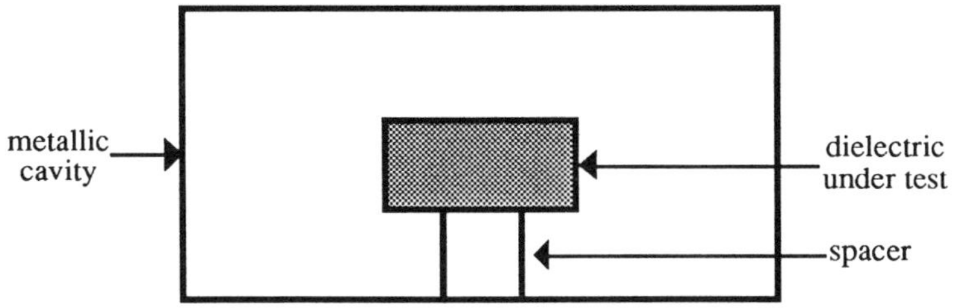

- Figure 3 -

Note that in this arrangement the material under test does not contact the metal surface of the fixture and thus there are no errors due to undesired gaps.

In the millimeter wavelength range quasi-optical open resonators [4] (figure 4) are substituted for metallic cavities. The sample is placed in a confocal Fabry-Perot resonator in a position where the electromagnetic fields are focused. Extremely high Q can be obtained with this set up allowing measurements of very low loss tangents.

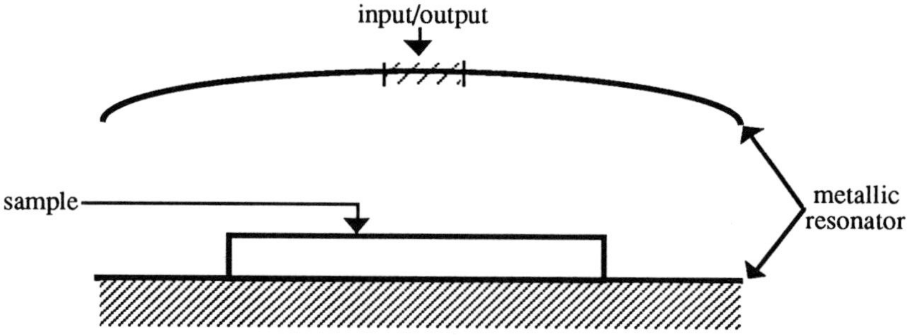

- **Figure 4** -

The resonant techniques are good for measuring very low loss materials and are capable of very high accuracy. However, it does involve rather complex analysis and since the material sample must be machined to a certain size and shape it is a destructive test. Also it is a single frequency measurement.

2 - The Reflection and Transmission Test Methods

They offer a means of measuring both the permittivity and permeability of a material over a very broad frequency band. They require the scattering parameters measurements of two port devices. The material is put into (or at the end) of a transmision line (or a waveguide) and the impedance of the air to material interface is measured. Depending on the techniques used, the sample holder can be a coaxial line (figures 5-a and b) or a metallic waveguide (figures 6-a and b) [5][6].

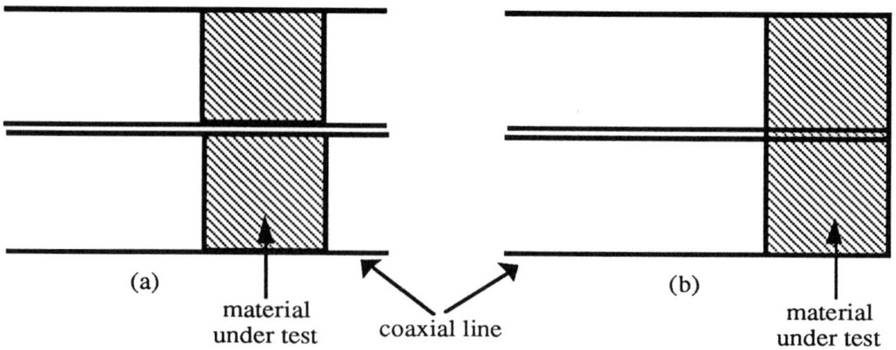

- **Figure 5** -

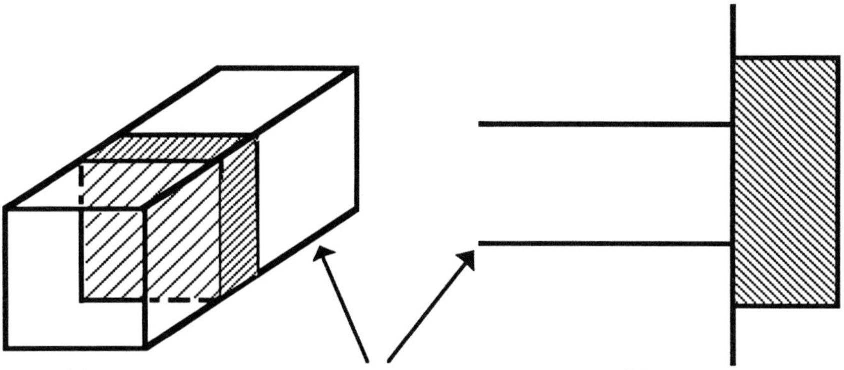

- Figure 6 -

Free space methods are inherently non destructive and the cross sectional shape of the sample is not important as long as it has flat parallel faces. Because the sample is not contacted by a fixture it is a good method for making high temperature measurements (figure 7) [7].

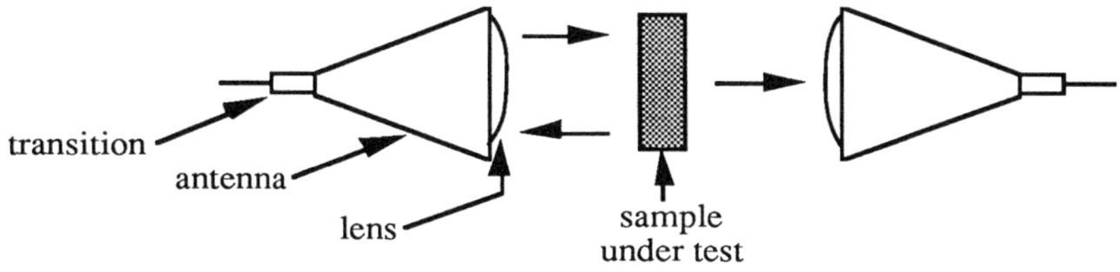

- Figure 7 -

In this free space method microwave energy is directed into the material sample with narrow beam antennas. It is important to focus the microwave energy into the center of the sample in order to avoid errors due to edge diffraction.

The scattering parameters equations are used to find the surface impedance and propagation constant of the material and they are solved for permittivity and permeability evaluation.

These reflection and transmission methods, which necessitate a calibration of the network analyzer, do not give accurate values of loss tangent for low loss materials, however they do give good results for the real part of permeability and permittivity. They are often used to evaluate electromagnetic parameters at high temperature. In this case, the sample under test is placed into a furnace [6][7].

SOURCES OF ERRORS

The major sources of errors in making material measurements are :

- instrumentation errors due to the network analyzer
- errors in the construction and application of the fixture

There are many errors associated with making a measurement with a modern network analyzer. In effect, directivity errors will affect the reflection coefficient measurement, while frequency response errors affect the measurement of the reflection and transmission coefficients as they vary with frequency. Note also that frequency accuracy is important in measuring both the resonant frequency and Q factor of cavity fixtures. Fixture errors can be due to fringing capacitance in transmission line method. Reflections from the fixture itself rather than the sample will also contribute to errors. Placement of the sample in the fixture is also important to make good measurements. Note also that some errors are associated with sample preparation and dimension measurements.

CONCLUSION

There are many different ways to measure the electrical properties of both dielectric and magnetic materials, but no universal measurement technique exists [8]. For example resonant methods work quite well for measuring low loss materials while free space methods are especially good for testing absorptive materials.

REFERENCES

[1] A. Julien, P. Guillon, "Complex permittivity and permeability measurements fo flat samples", *ISRAMT 91*, Reno, USA, 1991

[2] W.E. Courtney, "Analysis and evaluation of a method of measuring the complex permittivity and permeability of microwave insulators", *IEEE Trans. Microwave Theory Tech.*, Vol.MTT-18, pp.476-485, August 1970

[3] J.M. Durand - P. Guillon, "New method for complex permittivity measurement of dielectric materials", *Electronics Letters*, Vol.22, pp.63-65, January 1986

[4] T.B. Wells, R.L. Moore, D.C. Hicks, "Millimeter-wave materials measurement by Fabry-Pérot", *SPIC*, Millimeter wave technology II, Vol.544, pp.185-188, 1985

[5] N.E. Belhadj and al., "Broadband simultaneous measurements of complex permittivity and permeabiltiy using a coaxial discontinuity", *IEEE Trans. Microwave Theory Tech.*, Vol.MTT-38, pp.1-7, January 1990

[6] M. Chatard-Moulin, J.F. Villemazet, P. Guillon, H. Jallageas, "Non destructive complex permittivity measurements of absorbing materials", *IMTC 91*, Atlanta, USA, 1991

[7] V.V. Varadan, V.K. Varadan, "A free space method for measurement of dielectric constants and loss tangent at microwave frequencies", *IEEE Trans. Microwave Theory Tech.*, Vol.MTT-37, pp.789-793, June 1989

[8] P. Guillon, "Les actions de normalisation", *JCMM 1992*, Arcachon, France, 1992

CERAMIC-FILLED TRANSMISSION LINES FOR CIRCUIT MINIATURIZATION

John Deriso
Trans-Tech, Inc.
5520 Adamstown Road, Adamstown MD 21710

Abstract - **The high dielectric constants and low losses of modern ceramics enable circuit size reduction to meet miniaturization demands of new consumer products. The UHF band is traditionally awkward for realizing discrete inductors and capacitors, and air-filled transmission lines are impractically large. Metallized ceramics provide an attractive alternative. This overview of practical ceramic components highlights a few of the devices now appearing in wireless filter and oscillator circuits.**

INTRODUCTION

Decades ago, the 300 MHz - 3,000 MHz (UHF) frequency band was the domain of pioneering amateur radio enthusiasts. The tuned circuits needed at these higher frequencies required diminutive element values, and traditional coils and capacitors often became vanishingly small. Discrete components then gave way to distributed circuits, or transmission lines. Small size was not a prime concern, when portability implied transport in the trunk of an automobile. Personal communications devices were then fashion accessories in the minds of science fiction enthusiasts. But today's designers of consumer and military products for the UHF band are burdened with increasing demands for reduced product size and cost. The present "wireless communications revolution" forces a continuous tradeoff between performance and miniaturization for pagers, portable phones, televisions, and related products, which are now commonplace and some even on the threshold of wristwatch size.

COAXIAL LINES AS CIRCUIT ELEMENTS

A typical cordless telephone might require a tuned circuit at 931 MHz. An inductor-capacitor (L/C) combination of approximately 11 nanohenries and 3 picofarads would conceptually work, but the former would hardly be more than a 14 mm length of wire. The very connections to such components can result in parasitic inductances and capacitances which are large enough to de-tune the circuit. We call such tuned circuits *resonators*, because they store (electrical) energy and release it at a prescribed rate, much like the tone from a tuning fork or a bell.

Conventional Coaxial Cables

Coaxial lines are commonplace in every home equipped with cable television. These inexpensive cables are made with a soft, flexible *dielectric* (typically polytetrafluroethelyne) extruded over a stranded or solid wire *center conductor*, and covered with an *outer conductor*, or shield, constructed with braided wire or thin foil. The dielectric is chosen for its low signal loss over the considerable distance from the signal distribution box to the customer's television set. It's important to note that at UHF, electrical current travels very near the surface of the conductors. At 1000 MHz, for example, 95% of the RF current lies within 0.006 mm of the metal surface. This is commonly called the *skin effect*. A conventional coaxial cable, short-circuited at one end, would behave electrically similar to the 913 MHz tuned circuit, or resonator, if the cable were on the order of 6 cm long. Coaxial lines used as resonators are commonly cut to one quarter wavelength ($\lambda_g/4$, in the dielectric) at the operating frequency. But even this length is cumbersome to mount on the printed wiring board of a modern cordless phone. Modern ceramics provide the solution for miniaturization, since the length of a *ceramic-filled coaxial line* is reduced by the square root of the high dielectric constant.

Circuit Element Losses - Quality Factor

The *quality factor*, or "Q", of a UHF resonator is a measure of its ability to store the RF energy, or signal, without dissipating the signal as heat. Poor Q wastes transmitter power, or reduces receiver sensitivity, and either effect can result in less communication range and/or shorter battery life. Good Q is also essential for resonators in oscillators, where spectral purity is important in crowded communications bands. In general, a transmission line at UHF can provide an order of magnitude better Q than its discrete L/C circuit counterpart. Today's ceramic materials are of such high quality that the Q of ceramic-filled coax is limited primarily by resistive losses in the metallization which forms the conductors, rather than by losses in the ceramic.

CERAMIC-FILLED COAXIAL LINES - CONSTRUCTION

A short coaxial line - where flexibility is secondary to length reduction, and even undesirable for a stable tuned circuit - can be fabricated with a rigid ceramic dielectric. Unfired ceramic can be extruded in a tubular form, either circular or rectangular in cross-section. Once kiln-fired, the ceramic surfaces can be metallized within the center hole and around the outside of the extrusion, forming the center and outer conductors.

Choice Of Size, Dielectric Constant, and Temperature Coefficient

The ceramic is extruded in cross-sections ranging from 2 mm through 12 mm, and pressed in sizes through 19 mm. Size and Q are design tradeoffs available to the engineer: at a given frequency, the smaller the cross-section, the less the Q. Several ceramic compositions are offered, with relative dielectric constants (E_r) from 20 to 90. Popular ceramic compositions are barium nonatitanate, barium neodymium titanate, and magnesium titanate. The higher E_r results in 9:1 length reduction relative to an air dielectric, which is an advantage at the lower end of the UHF range where wavelengths are long. In the 1000 MHz to 3000 MHz region, intermediate E_r on the order of 36 is popular for a 6:1 length reduction. An E_r of 20 is used at the highest frequencies and where lowest losses are necessary, for Q also tends to increase as E_r decreases. The Q of a dielectric-filled coaxial line may be expressed as

$$\frac{1}{Q} = \frac{1}{Q_c} + \frac{1}{Q_d} \tag{1}$$

where Q_c and Q_d represent the conductor and dielectric losses, respectively. Typical values for a 4 mm cross-section coaxial line, with E_r of 87 at 1000 MHz would be

$$Q_c = 350 \text{ (thick-film silver)}$$
$$Q_d = 4500$$
$$Q = 325$$

As mentioned earlier, conductor losses dominate the overall quality factor.

The *temperature coefficient of resonant frequency* (τ_f) of a tuned circuit is a measure of its ability to maintain frequency as temperature varies. Expressed in parts per million per degree Centigrade (ppm/°C), τ_f ranges typically from -15 ppm/°C to +6 ppm/°C for popular ceramic compositions. For a given material, τ_f

is essentially constant within the operating temperature range of low-cost, commercial electronic products. Zero τ_f is ideal for inexpensive, fixed-frequency filters. Free-running oscillators, with frequency dependent upon the resonator, may actually benefit from non-zero τ_f for temperature compensation.

Metallization

Usually, high-conductivity silver is thinned for spray application, but gold or copper is also applied. The film metallization - typically 0.013 mm thick - is oven-fired to insure adhesion to the ceramic. Skin effect at UHF permits a thin, hollow center conductor and thin outer shield.

Mounting And Circuit Connections

A stamped metal tab is usually inserted and soldered into the metallized center conductor hole at one end to provide circuit connection. Typically mounted on a printed wiring board, the ceramic coaxial line may be tape-and-reel packaged for pick-and -place location, surface mounted on its side, and soldered directly to the foil on the board.

QUARTER-WAVE RESONATORS - CERAMIC CIRCUIT BUILDING BLOCKS

Once the dielectric constant has been accurately determined for a batch of ceramic, the extrusions may be sawn to precise dimensions which are quarter-wavelength ($\lambda_g/4$) in the dielectric at a specified frequency. Length accuracy of ±.025 mm is commonplace. The $\lambda_g/4$ length has interesting properties, depending on whether the line is shorted or open at the far end. In the former case, the end of the coaxial line without the contact tab is metallized, short-circuiting the inner and outer conductors at that point. To an RF signal, the shorted line appears (resonates) similar to a *parallel-tuned* L-C tank circuit. When open-circuited at both ends, a quarter-wavelength coaxial line behaves electrically (resonates) similar to a *series-tuned* L-C circuit, sometimes referred to as a trap. The equivalence of quarter-wave and half-wave coaxial lines to lumped (L/C) tuned circuits is described in the literature [1].

Table 1 shows some typical examples of quarter-wave resonators, and indicates the tradeoffs which can be made between size reduction and Q factor.

UHF FILTERS WITH CERAMIC-FILLED COAXIAL LINES

Resonators formed from coaxial lines may be used individually or combined

TABLE I
Examples of $\lambda_g/4$ Resonators

E_r	τ_f ppm/°C	Resonator Length $\lambda_g/4$ @ 1000 MHz	Cross-Section	Q
1.00 (air)	-17*	74.9 mm*	12 mm*	1600
20.6	0 ± 5	16.5 mm	12 mm	1200
38.6	5 ± 1	12.1 mm	12 mm	980
			6 mm	577
			4 mm	427
87	-13 ± 1	8.0 mm	6 mm	480
			4 mm	322

* hypothetical coaxial line fabricated from copper conductors and air dielectric

in various ways to provide selective filtering of UHF signals. Several types of inexpensive filters are now commercially available. Ceramic filters offer low insertion loss, small size, excellent thermal and mechanical stability, and low cost.

Bandpass Filters

Consumer products frequently use narrow-band filters as circuit building blocks. Emphasis is typically upon low cost and size, currently on the order of $2.00 or less per resonator and approximately 2 cm^3 for a 3-resonator cordless phone filter. Several lines are typically coupled together to enhance the selectivity of the bandpass characteristic. Coupling between resonators can be achieved with small external capacitors, or more commonly, by apertures formed in the outer metallized walls between resonators. Bandwidths on the order of a few per cent of center frequency are practical. At least one designer [2] claims to adjust the individual resonators and assemble a filter with no subsequent tuning. More often, the resonators are pre-cut slightly below the desired center frequency of the filter, assembled, and then the overall filter response is fine-tuned by selectively abrading away small areas of the outer metallization of each resonator.

Power handling capability is limited by the relatively poor thermal conductivity of ceramic, and by the resonator Q, which governs the heat generated from signal loss. The problem is getting the heat out without causing a large thermal gradient and differential expansion which might crack the ceramic. Still, bandpass filters can handle power levels on the order of 1 watt/cm^3, quite enough for typical cellular phones. Higher power levels are accommodated by designs with 12 mm or larger cross-section resonators.

The temperature stability of ceramic filters can be quite good. Five-resonator 1.6 GHZ filters have been fabricated from ceramic with $E_r = 20$ material, operating with less than 1 MHz drift over the -55°C to +95°C temperature range.

Table II shows a comparison of ceramic filter attributes compared to competing technologies such as discrete-component and surface acoustic wave (SAW) filters.

Band Reject Filters

Occasionally an interfering signal, such as a weather radar, needs additional suppression in a receiver. Band-reject filters attenuate specific frequencies and pass others with low loss. As in the case of bandpass filters, several resonators can be combined [3] to enhance the selectivity. When coupled with quarter-wavelength (typically microstrip) lines on a substrate, narrow to moderate rejection bandwidths can be achieved.

Diplexers

Portable cellular phones require a means to connect the transmitter and receiver to a common antenna, while isolating the circuits from each other. When the transmit and receive frequencies differ, filters can be connected, or combined to perform this *diplexing* function. Depending upon the application, the filters may be two discrete bandpass filters, two discrete band reject filters, or one of each. The combining is typically done on the printed wiring board, with a judicious choice of circuit traces joining the filter inputs to the common antenna connection.

Notch Filters

A notch filter is a special class of band-reject filter, often fabricated with a single ceramic resonator, near half-wavelength long and *open* circuit at each end.

Table II. Bandpass Filters - Competing Technologies

	Ceramic	Air Dielectric	Discrete Component	Helical	Microstrip	SAW
Insertion Loss	Low 0.3 - 5.0 dB	Low 0.3 to 5.0 dB	Medium 3.0 to 10.0 dB	Medium 3.0 to 10.0 dB	Medium 3.0 to 10.0 dB	Medium/High 3.0 to 20.0 dB
Power Handling	Medium 1 to 50 Watts	High > 100 Watts	Medium < 50 Watts	Low to Medium	Low to Medium	Low <10 dBm
Physical Size	Small	Large	Small to Medium	Medium	Thin Medium Surface Area	Very Small
Temperature Stability	Very Good	Poor	Poor to Fair	Poor to Fair	Fair	Very Good
Mechanical Stability	Very Good	Poor	Poor to Fair	Poor to Fair	Very Good	Very Good

When connected in shunt (to ground) through a small capacitor to a printed wiring board, the resonator will trap a narrow band of frequencies. The width of the notch may be adjusted by altering the capacitor value, and with clever selective metallization, the capacitor may even be integrated into the same coaxial ceramic line.

Highpass Filters

Rectangular waveguides can be filled with dielectric to reduce their cross-section and length. Though not coaxial, they are commonly used as transmission lines, and can pass signals having wavelengths less than a critical cutoff wavelength, which is determined by the guide width and the E_r filling the waveguide. Frequencies higher than that corresponding to the cutoff wavelength will pass, while lower frequencies will be attenuated. In the latter region, the length of the waveguide determines the degree of attenuation.

A rectangular ceramic bar is formed of high-E_r material and coated on the long four sides with thick-film metallization. Signals are introduced into one end of the waveguide, typically with ribbon connections to the printed wiring board, and are similarly recovered at the far end of the waveguide. An example highpass filter would be 7.5 mm X 2.5 mm X 27.1 mm, consisting of thick-film silver-metallized ceramic with E_r of 90. This filter would pass signals above 2100 MHz, and attenuate signals below 2000 MHz by more than 30 dB.

UHF OSCILLATORS WITH CERAMIC-FILLED COAXIAL LINES

Oscillators require high-Q resonators for stability and lowest phase noise, or spectral purity [4]. Voltage-controlled oscillators (VCO's) require an equivalent inductance to resonate with the tuning varactor. Here, coaxial line elements can simulate high-Q, temperature-stable, compact inductors. More precisely, shorted coaxial lines will exhibit an inductive reactance when used below quarter-wave resonance, and will approximate the behavior of an ideal inductance or "coil" over a limited frequency range. Ceramic resonators are appropriate because they minimize *microphonics*, or frequency instability caused by mechanical motion. An unshielded resonator such as a microstrip line may be itself mechanically rigid, but could be influenced by the motion of a vibrating cover or by the proximity of a human hand. The shielded, compact size of ceramic lines affords good temperature stability and compatibility with pick-and-place assembly. The schematic of Figure 1 shows the electrical location of a coaxial resonator in a typical VCO circuit.

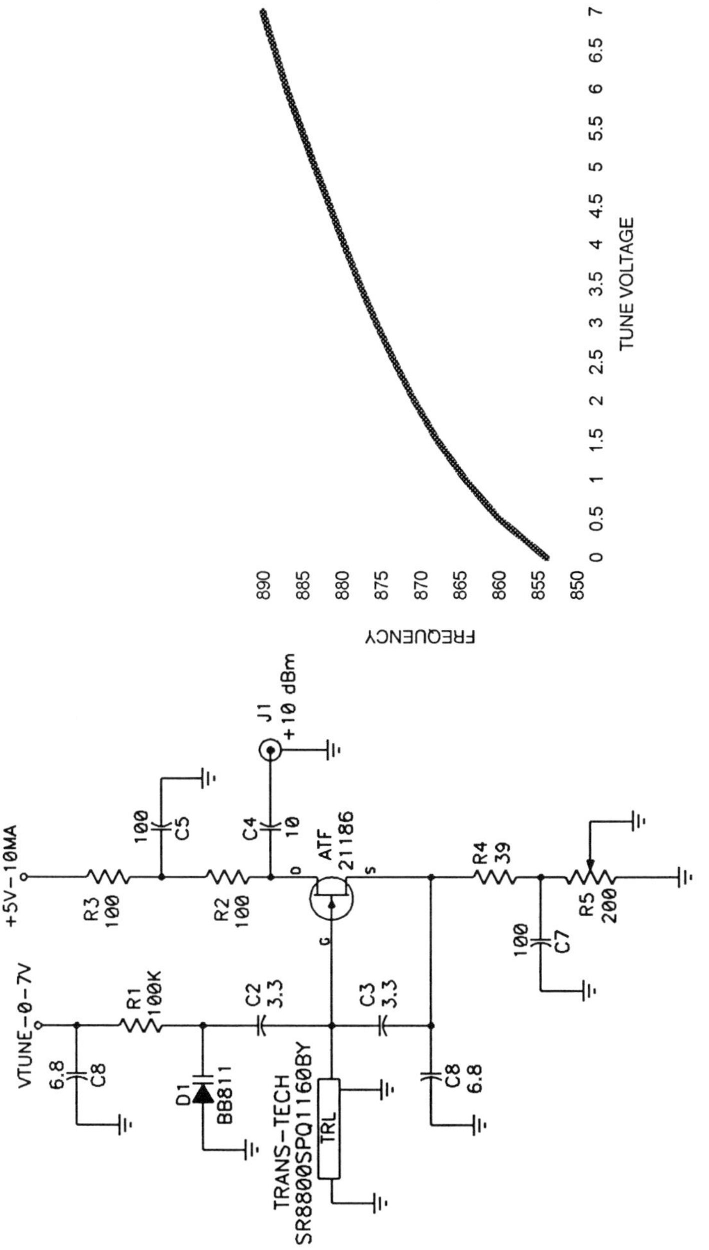

Figure 1. Typical VCO Circuit Schematic Showing Coaxial Resonator (TRL) and Tuning Curve

CONCLUSION

Circuit design in the UHF band was once an ordeal of choice between tiny discrete components and awkwardly large distributed transmission-line circuits. Temperature-stable ceramic materials with high dielectric constants have reduced the size of transmission lines to manageable, economic proportions and have breathed new life into older, established UHF design techniques.

REFERENCES

[1] T. Moreno, "Microwave Transmission Design Data," Artech House Books, Norwood, MA, 1989, p. 40.

[2] K. Hano and K. Sawamoto, "A Bandpass Filter Using Direct-Coupled Quarter-Wavelength Coaxial Dielectric Resonators," Microwave Journal, November 1987 pp. 141-160.

[3] L. Young, G. Matthaei, and E. Jones, "Microwave Band-Stop Filters With Narrow Stop Bands," IRE Trans. Microwave Theory Tech., November 1962, pp. 416-427.

[4] U. Rohde, "Oscillator Design for Lowest Phase Noise," Microwave Engineering Europe, May 1994, pp. 31-40.

Designs and Materials

NUMERICAL SIMULATION OF ELECTRICAL CHARACTERISTICS OF MULTILAYER CERAMIC CAPACITORS

P.K. Ghosh and M.E. Azimi,
Department of Electrical and Computer Engineering
Syracuse University
Link Hall
Syracuse, NY 13244-1240

ABSTRACT

Multilayer ceramic capacitors (MLCs), with their inherent volumetric efficiency, are considered to be one of the key components of the modern electronic circuits. The internal and the external metallic electrodes of MLCs are prone to damage during the fabrication and/or the assembly process. In addition there are possibilities of microstructural modulation of the host dielectric. These changes can result in an electrical response which significantly differs from the expected one. This paper presents a theoretical approach which enables simulation of the electrical environment inside the multilayer ceramic capacitors as a function of the embedded electrode's shape and the position. This approach is general in nature and allows calculation of the electric field, the potential distribution and the effective capacitance of the multilayer structure. Specific structures analyzed with numerical calculations will be presented along with the results obtained using another well known analytical approach.

INTRODUCTION

In recent years multilayer ceramic capacitors (MLC) and actuators (MLCAs), with similar structure as MLCs, have been developed quite extensively [1]. Fig. 1 shows schematic of a typical multilayer structure. For enhancing the understanding of the multilayer structure, both for reliability and future development, it is very important to perform rather thorough analysis of the electrical environment within the multilayer structure.

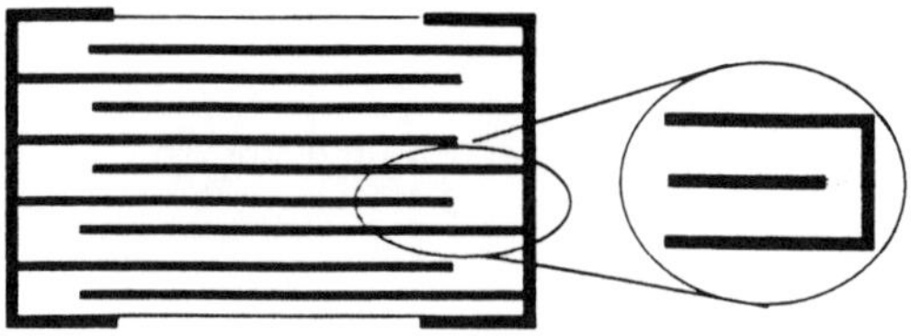

Fig. 1 A representative ceramic multilayer structure.

Many analyses are simplified with the assumption of a homogenous ceramic dielectric. However, after processing the resultant ceramic often turns out to be a rather heterogeneous system with embedded flaws (Fig. 2). Voids, pores, delamination, misplaced and/or rough electrodes are some of the common defects of this structure that are produced due to the inconsistencies in the fabrication process. Mechanical and thermal shocks during packaging and assembling on the circuit board bring additional problems. These changes can result in an electrical response which significantly differs from the expected one.

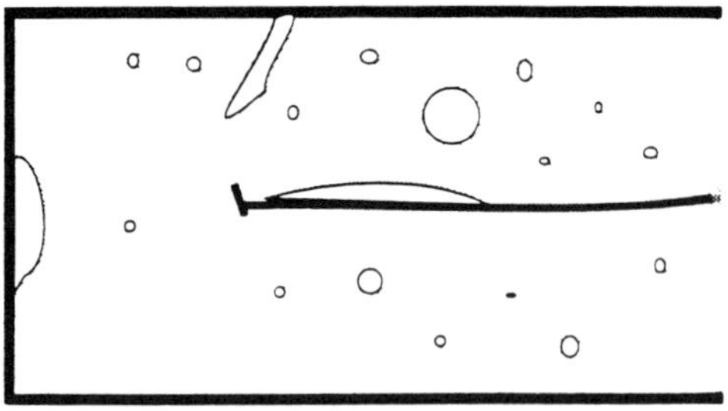

Fig. 2 Processing flaws within ceramic multilayer structure.

In this paper we report a finite element method to simulate the electrical environment inside the flawed multilayer structure. Our results explicitly show the changes in electric field distribution due to the presence of voids, and position and roughness of the buried electrode. Also we calculate the effect of these imperfections on the capacitance value of the multilayer structure. This information can be used to predict some of the possible failure mechanisms in modern multilayer components.

FINITE ELEMENT METHOD

The finite element method that we will use here is based on the approximate solution of the variational form of the Poisson's equation as follows [2]:

$$-\nabla \cdot \left(\bar{\bar{\varepsilon}} \nabla \psi\right) = 4\pi\rho \tag{1}$$

where $\bar{\bar{\varepsilon}}$ is the permittivity tensor, ρ is the source charge density and ψ is the potential.

For two-dimensional cases the above equation reduces to

$$-\frac{\partial}{\partial x}\left(\varepsilon_x \frac{\partial \psi}{\partial x}\right) - \frac{\partial}{\partial y}\left(\varepsilon_y \frac{\partial \psi}{\partial y}\right) = 4\pi\rho \tag{2}$$

where ε_x and ε_y are the components of the permittivity tensor $\bar{\bar{\varepsilon}}$. Also, both permittivity and charge density could be functions of space variables x and y.

Next, to derive the variational form of the above equation (eq. 2) we first multiply each term by a sufficiently smooth test function $\phi(x,y)$ and then integrate over the whole domain [3]. Now, let Ω represent the whole domain of the problem and Ω' is the domain of an element. For our calculation we took triangular elements. First, we seek solution on a representative triangular element. Next, applying the continuity condition on the resultant solution (or its derivatives) on the common edges of triangular elements, we assemble all elements to find the solution for the whole domain Ω.

Now, assume permittivity tensor $\bar{\bar{\varepsilon}}$ is constant over each element. Then by applying Green's theorem [4] the following relationship can be derived for each element

$$\int_{\Omega'} \left[\varepsilon_x \frac{\partial \psi}{\partial x} \cdot \frac{\partial \phi}{\partial x} + \varepsilon_y \frac{\partial \psi}{\partial y} \cdot \frac{\partial \phi}{\partial y} \right] ds = \int_{\Omega'} \rho \phi ds + \int_{\Gamma'} \left[\varepsilon_x \frac{\partial \psi}{\partial x} \cdot n_x + \varepsilon_y \frac{\partial \psi}{\partial y} \cdot n_y \right] ds$$

...(3)

where Γ' denotes the boundary of an element, n_x and n_y are the components of the unit normal vector on the boundary Γ' at point (x,y).

We now introduce the finite element approximation of the form

$$\psi(x,y) = \sum_{j=1}^{m} \psi_j \omega_j(x,y) \qquad (4)$$

where m is the number of nodal points and ψ_j is the nodal value of the potential at the jth node. $\omega_j(x,y)$ represents the finite element shape function.

Substituting equation (4) in equation (3) yields a matrix equation, in the following form, for each element

$$[a_{ij}] \; [\psi] = [b] \qquad (5)$$

where $[a_{ij}]$ is a square matrix, and $[\psi]$ and [b] are column matrices. Elements of matrix $[a_{ij}]$ and vector [b] can be calculated by solving the problem on each element separately and then assembling the resulting micromatrices.

For simplicity we have assumed that the potential ψ over each element varies linearly. Thus the shape function $\omega_j(x,y)$ is also a linear function over each element. Also, the permittivity tensor for the ith element is given by

$$\bar{\bar{\varepsilon}}^i = \begin{bmatrix} \varepsilon^i{}_x & 0 \\ 0 & \varepsilon^i{}_y \end{bmatrix} \qquad (6)$$

For a linear shape function, the 3×3 matrix for each element will assume the form of

$$\frac{1}{2D}\begin{bmatrix} \varepsilon^i_x q_i q_i + \varepsilon^i_y \gamma_i \gamma_i & \varepsilon^i_x q_i q_j + \varepsilon^i_y \gamma_i \gamma_j & \varepsilon^i_x q_i q_k + \varepsilon^i_y \gamma_i \gamma_k \\ & \varepsilon^i_x q_j q_j + \varepsilon^i_y \gamma_j \gamma_j & \varepsilon^i_x q_j q_k + \varepsilon^i_y \gamma_j \gamma_k \\ \text{symm.} & & \varepsilon^i_x q_k q_k + \varepsilon^i_y \gamma_k \gamma_k \end{bmatrix}$$

where $q_i=y_j-y_k$, $\gamma_i=x_k-x_j$. D is the area of the element and i,j,k are the indices for three vertices of the triangular element. The assembling of these micromatrices and imposing the boundary conditions yield the final matrix equation which must be solved for global finite element problems.

In the present investigation, solution of the resultant M×M matrix equation yielded the potential value at each node within the global domain Ω. Electric field distribution can be subsequently calculated with little effort. True capacitance value between two electrodes with defects within the structure is calculated by first estimating the total charge on each electrode. Total charge (Q) on an electrode can be expressed as

$$Q = \int_{C_j} \varepsilon(x,y) \nabla\phi \cdot \bar{n}\, d\ell \qquad (7)$$

where C_j is any closed contour enclosing one of the electrodes, $\bar{n}$ is the normal on the contour, and $d\ell$ is the differential element along the path of contour.

The effective capacitance (C_{eff}) is then calculated from the applied voltage across the two electrodes under consideration.

SIMULATION RESULTS

We have developed computer software for analyzing the electrical properties of a multilayer system with different types of flaws. Fig. 3 shows one such section, and also includes the mesh arrangement used for calculation. Figs. 4 and 5 show the potential and electric field distribution when the voids are present within the structure. Permittivity of the void is much less than that of the host material. Significant change in potential distribution is observed around the voids. Fig. 6 shows the magnitude of the electric field around the voids.

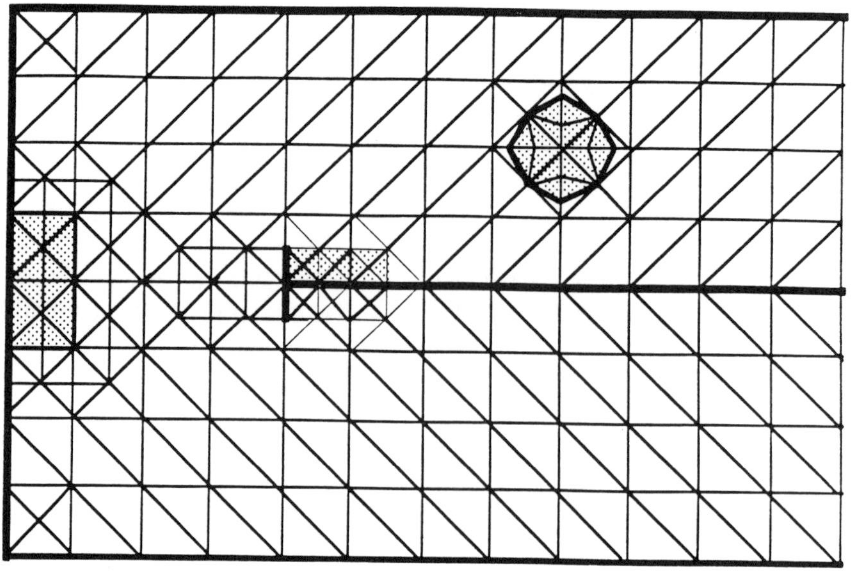

Fig. 3 A section of the multilayer structure along with the mesh for simulation.

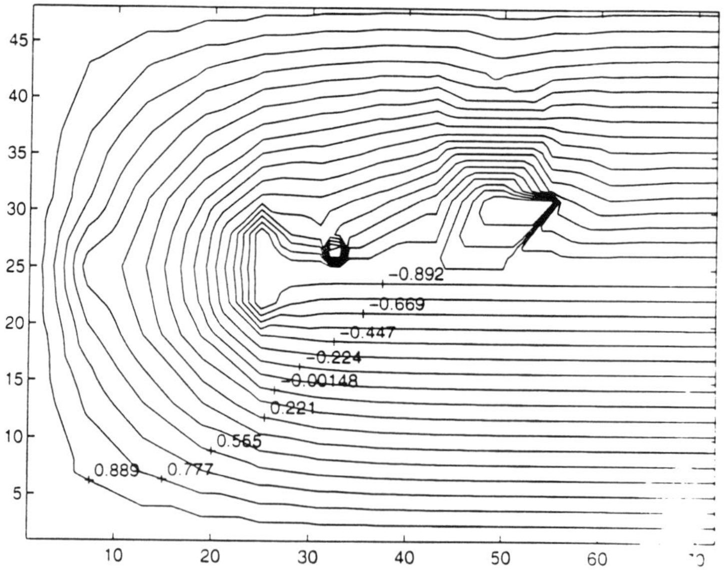

Fig. 4 Equipotential line inside the structure with flaws.

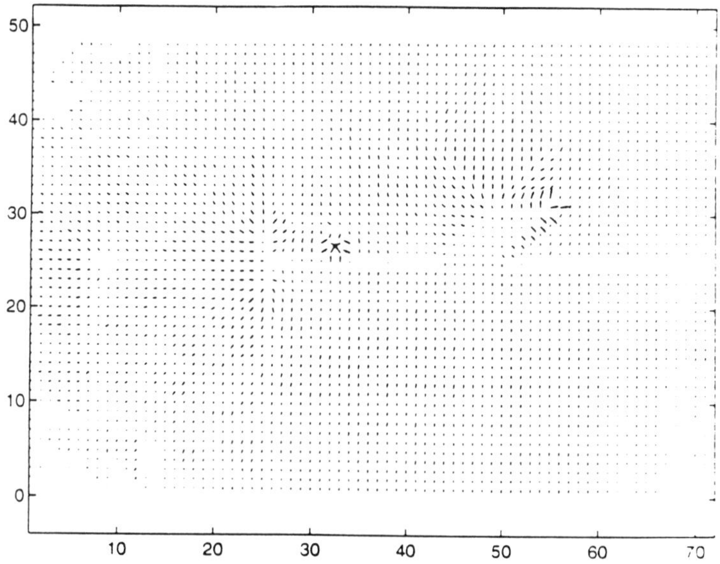

Fig. 5 Electric field distribution inside the structure with flaws.

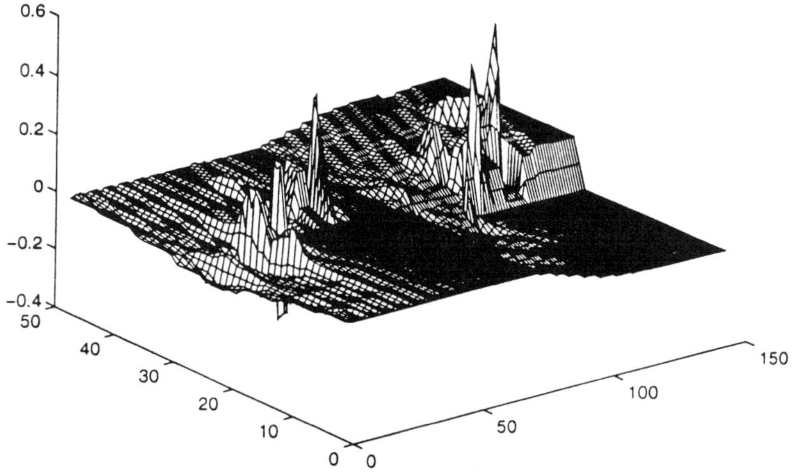

Fig. 6 Magnitude of the electric field.

If now the flaws are created by inclusions with permittivity higher than the host material, then the potential and electric field distribution are expected to be opposite of the previous case, and are confirmed in Figs. 7 and 8.

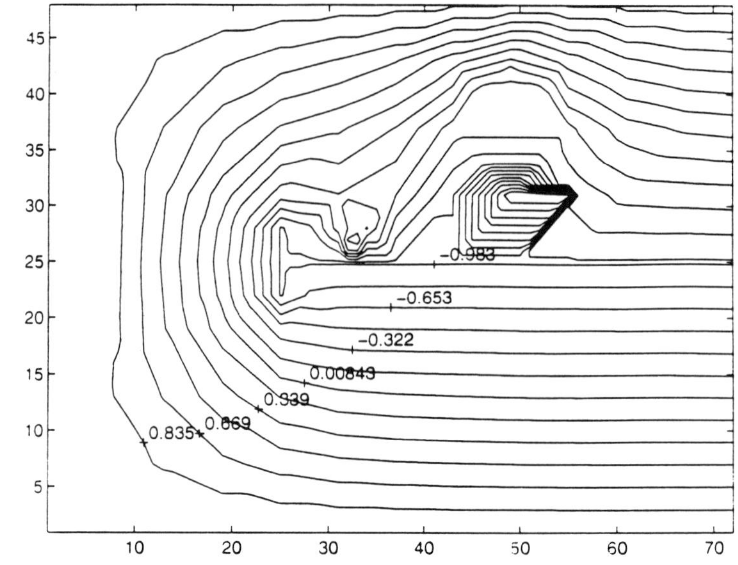

Fig. 7 Equipotential lines when the permittivity of inclusion is higher than the host ceramics.

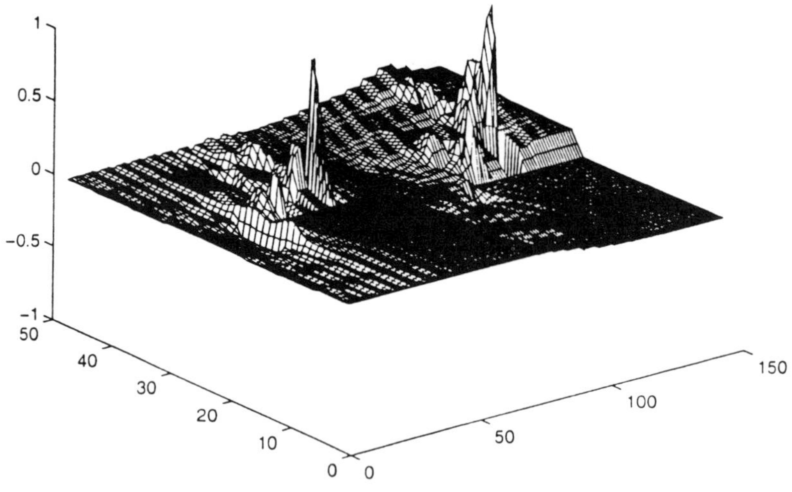

Fig. 8 Magnitude of the electric field corresponding to Fig. 7.

Figs. 9 and 10 show the effect of anisotropic dielectric property of the host material where $\varepsilon_y > \varepsilon_x$. We observe a high electric field along the low permittivity direction. Therefore, for a poled ceramic, possible failure and fracture propagation will be more likely along the unpoled low permittivity direction. This phenomenon has been observed experimentally [5].

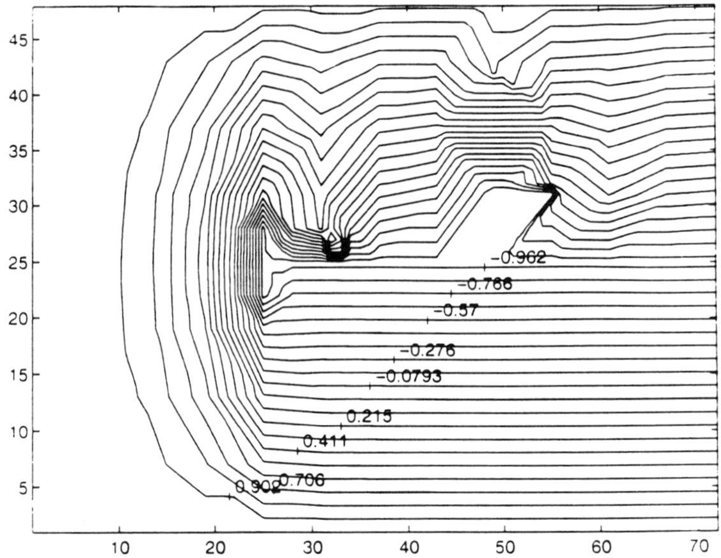

Fig. 9 Effect of Anisotropic dielectric property of the host material on the potential distribution.

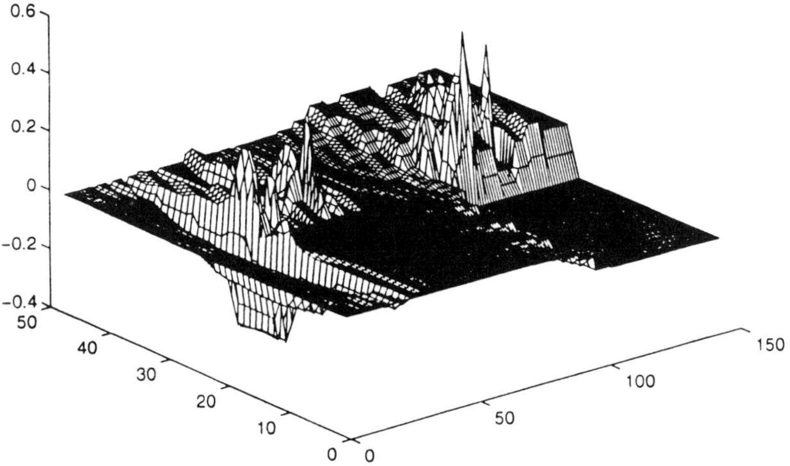

Fig. 10 Effect of the Anisotropy on the electric field.

Finally, a simple parallel plate configuration may not always accurately estimate the capacitance value for the multilayer structure. Our calculations, as shown in Fig. 11, show that the capacitance value of a multilayer structure depends on the volume fractions of voids. It is to be noted that the effects of other parameters such as the electrode aspect ratio and number of layers can also be accurately simulated using this method.

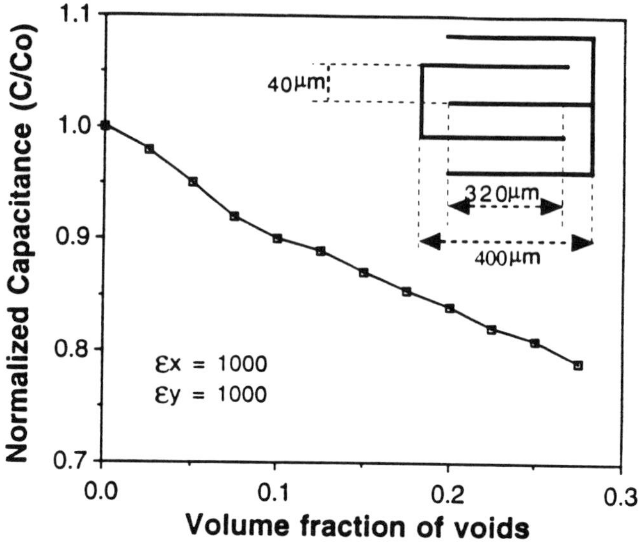

Fig. 11 Calculated normalized capacitance of the multilayer structure. C_0 represents capacitance value with no flaws.

In conclusion, we have presented a general approach to map the electrical environment within multilayer ceramics. This method supplements the investigations of other researchers. We have been successful in identifying the nature and the extent of effects of various flaws on the electrical behavior of different host materials. This method provides an accurate and user-friendly tool for the design and analysis of multilayer ceramic structures.

REFERENCES

[1] Proceedings of the Center for dielectric studies symposium on improvement of multilayer ceramic reliability, ACers and ISHM, May 1991.

[2] J.A. Stratton, "Electromagnetic Theory," McGraw-Hill, New York, 1941.

[3] Claes Johnson, "Numerical solution of partial differential equations by the finite element method," Cambridge, Sweden, 1992.

[4] Ivar Stakgold, "Green's functions and boundary value problems," Wiley, New York, 1979.

[5] T. Yamamoto, H. Sato, H. Igarashi and K. Okazaki, "Stress anisotropy of PLZT ceramics induced by polarization," Jap. J. of Appl. Phys., Vol. 22, Suppl. 22-2, pp. 70-72, 1983.

EFFECT OF DOPANTS ON HIGH FREQUENCY DIELECTRIC PROPERTIES OF LEAD ZIRCONATE

Rong-Fong Huang and E. S. Ramakrishnan
Ceramic Technologies Research Laboratory,
Land Mobile Products Sector, Motorola , Albuquerque, NM 87113

ABSTRACT

The effect of oxide additives on the high frequency dielectric properties of lead zirconate materials are studied. The dielectric loss of modified $PbZrO_3$ compositions has been improved by the addition of small amounts of manganese oxide using conventional sintering method in the temperature range of 1200-1280°C. A composite stripline resonator suitable for high frequency applications has been prepared using this compound and a positive T_f ceramic to achieve temperature compensation.

INTRODUCTION

In the last several years there has been a significant interest and research activity on the development of high frequency dielectric materials [1] for telecommunication applications. For these applications, the dielectric must exhibit high dielectric constant, K, low loss (high Q) and possess low temperature coefficient of resonant frequency (T_f). High Q dielectric materials with K in the range of 20-90, Q of about 3000 to 30000, and temperature coefficient in the ranges of 0 to 60 ppm/°C, have found wide applications for microwave components such as resonators, filters etc. One common approach to reduce the T_f of the dielectric materials is to formulate the compound by mixing a positive T_f material with a negative T_f material. Using this temperature compensation method low T_f, high Q dielectrics have been successfully developed[2]. However, many mixed phase materials tend to react with each other during sintering and result in poor Q or low K for the mixed compositions. A second approach of physically bonding the two separately fired ceramics with opposite signs of T_f, has also been studied previously [3,4]. This method prevents the reaction problems encountered in the mixed phase approach.

The thrust towards miniaturizing device sizes necessitates the development of even higher K materials [5], since the size of the component is inversely proportional to $K^{1/2}$ of the material. Dielectrics such as $CaTiO_3$ and $SrTiO_3$ based ceramics, are known to exhibit K in the range of 150 to 350 with low loss, but they also exhibit large positive T_f which limits their usefulness for high frequency applications. Thus, for these high K materials to be useful, their T_f needs to be compensated by compounds with opposite sign of T_f.

Since most known high K dielectrics show a large positive T_f, they require materials with large negative T_f to compensate. There are only a few dielectrics that exhibit negative T_f with high K and high Q at high frequencies. One such material with a large negative T_f and a high dielectric constant is lead zirconate ($PbZrO_3$). It is an anti-ferroelectric material that has a relatively high K of about 120, as compared with other paraelectric dielectrics at microwave frequencies. But the Q of $PbZrO_3$ is only 40 at 500 MHz when prepared by the conventional sintering methods [6]. Tatsuki et al [3] have studied the effect of rare earth oxides as sintering aids and found Gd, Ce, and Tb are effective additives to raise the Q. They achieved a Q of 850 when samples with these additives were prepared using hot-press method.

In this work, we have studied methods to reduce the high frequency loss of lead zirconate using the conventional sintering techniques. The effect of different oxides additives for $PbZrO_3$ was investigated in order to reduce the dielectric loss. Compositions modified with the addition of both MnO and CeO_2 were studied in detail and the high frequency properties of the materials were measured. The effect of sintering temperature on the Q of the sample was also investigated. Composite stripline resonators were fabricated using substrates prepared from modified $PbZrO_3$. These striplines were constructed by sandwiching the negative T_f substrate ($PbZrO_3$) with a positive T_f substrate made by $(Sr,Ca)TiO_3$. The T_f of the composite structure was measured to determine the effectiveness of temperature compensation by the modified $PbZrO_3$.

EXPERIMENTAL

The samples were prepared from PbO and ZrO_2 as the starting materials with or without a small amount of CeO_2 incorporated as a grain size modifier. When searching for suitable additives, a series of 2 mol% CeO_2 doped $PbZrO_3$ powders with oxide additives (CaO, SrO, BaO, MnO, MnO_2, Bi_2O_3, Nb_2O_5, Ta_2O_5, WO_3) ranging from 0 to 20 mol% were prepared. An excess PbO was added to the composition to compensate for the lead loss during the sintering. Conventional ceramic processing techniques were used to make the powder. The mixed powders were calcined at 750-900°C for 1-2 hours and the calcined material was ball milled, dried, pulverized, sieved and mixed with binder. The powder was pressed into discs, blocks or cylindrical resonators. The substrate samples were prepared using the tape casting method with appropriate binder system. After subjecting to a slow debinderization step, the samples were sintered in ambient atmosphere from 1100-1300°C for different lengths of time in a lead-enriched enclosure to minimize inhomogeneities in the composition caused by lead volatilization. The cooling cycle was programmed, taking into consideration the phase transformation at ~230°C, to prevent the development of microcracks.

An LZC meter (HP4276) was used to measure low frequency dielectric properties (1KHz-20KHz) and the temperature coefficient of the dielectric constant (T_K). The high frequency dielectric constant and loss factor were measured using Boonton Model 34A resonant coaxial transmission line [7] coupled to Model 102D Signal Generator and Model 92B RF Millivoltmeter. The chip samples for the Boonton measurements were prepared from the sintered discs/blocks with fired-on silver as electrodes. The sample thickness was adjusted to keep the capacitance to be in the range of 5-10 pF so that the measurement frequency is maintained in the range of 450±25 MHz. Resonator or stripline samples were prepared and placed into a computer controlled oven (Delta Design) for T_f measurements. A network analyzer (HP 8753A) was used to measure the resonance frequency shift of the sample caused by the temperature variation.

RESULTS AND DISCUSSION

(1) Effect of oxide additives on Q:

The Q of undoped $PbZrO_3$ (PZ) has been reported to be 40 at around 500MHz by Lanagan et al [6] when prepared by regular pressureless sintering. In this study, the high frequency properties of PbO-ZrO_2-CeO_2 (PZC) ceramics were studied using the similar sintering method. Our samples showed K of ~110 and Q of ~300 at high frequencies when sintered at 1200°C for 1-2 hours. These results confirm the effect of CeO_2 in enhancing the Q of PZ. But the Q is still much lower than that achieved by the hot press method. Tatsuki et al [3,4] had found that the Q of PZ can be raised up to 850, using a hot press method, with 2 mol% of CeO_2 addition as sintering aid. It is believed that by using suitable dopants and sintering aids, higher Q can be achieved by the conventional sintering techniques. Since CeO_2 was effective to raise the Q, it was added as part of the base formulation for this study. Several oxides were chosen to modify the PZC composition to determine which additives can be used to improve the dielectric properties. Results of these samples fired for one hour at 1200°C are listed in Table I, where the K's and Q's are measured at 450±25 MHz and the T_K at 20 KHz. For the CaO, BaO, SrO, MnO or Bi_2O_3 additions, the Q generally showed moderate improvements over the samples with no additives and little change on K and T_K were seen on samples up to 2.5 mol% additions. On the other hand, the addition of Nd_2O_5, Ta_2O_5, and WO_3 resulted in lower Q and higher T_K for PZC. From these results, it is evident that the addition of lower valence cations can enhance the Q of PZC whereas cations with higher valence states (+5 or +6) have an adverse effect. Similar additive effects were observed for other lead containing compounds.[8] The reduction of dielectric loss by acceptor doping was attributed to the formation of oxygen vacancies to compensate for the divalent cations additions. The pinning of oxygen vacancy reduces the domain wall mobility, hence can increase the dielectric Q of domain containing compounds. On the other hand, high valence cations tend to facilitate the domain wall motion and result in high dielectric loss factor. Since PZ is an antiferroelectric compound which also exhibits domain walls, it is reasonable that the effect of additives on dielectric loss shown in Table I can be attributed to the domain wall mobility as well. However, other more important effect, to raise the Q of PZC, were seen on MnO doped samples which will be discussed in more detail in the following section.

TABLE I. Effect of 2.5 mol% Addition on the Dielectric Properties of $Pb_{1.06}Zr_{0.98}Ce_{0.02}O_3$ Sintered at 1200°C for 1 hour

Additive	Q	K	T_K (ppm/°C)
None	300	109	2250
CaO	331	93	2268
SrO	314	103	2836
BaO	361	112	
MnO	396	110	2478
Bi_2O_3	268	105	2875
Nb_2O_5	14	171	
Ta_2O_5	29	55	4100
WO_3	207	134	3035

Of all the additives studied, the addition of MnO was found to be the most effective to reduce the loss without significantly affecting the K. Though Mn did cause an increase in the T_K, the value of which was within the range of compensation by other materials in a composite structure. The Mn added samples were studied in detail to find out the optimum additive levels as well as firing profiles.

(2) Effect of Mn addition on PZC properties

Samples with MnO contents ranging from 0.25 to 5 mol% were prepared using several firing profiles and their properties measured. Figure 1 shows the measured Q values for samples sintered at 1200°C with different amount of manganese additions to PZC. As can be seen, mixing the PZC with 0.25-1.0 mol% MnO results in a substantial improvement of Q. An addition of 0.25 mol% is seen to increase the Q from about 300 to 575 at around 450 MHz. However, further Mn addition of amounts more than 1 mol% has a detrimental effect on the Q. At 5 mol% Mn, the Q drops to values lower than that of the original composition. It is interesting to note that the Q values do not follow the densities monotonously.

Although the addition of MnO shows a distinct effect on Q, it does not seem to affect the K very much. As shown in Figure 2, the K increases slightly with respect to the MnO content. This increase can be attributed to the density increase seen in MnO containing samples shown in Figure 1. The T_K also shows a trend in Figure 2. It increases slightly with the incorporation of MnO up to 2.5 mol%, above which, the T_K increases drastically. This behavior is similar to that reported for the Mn doped $Ba(Mg_{1/3}Ta_{2/3})O_3$ [9]. Since the T_k is related to T_f by the relation of $T_f = -0.5T_K - TEC$, where TEC is the thermal expansion coefficient, the T_f is expected to increase with the MnO additions.

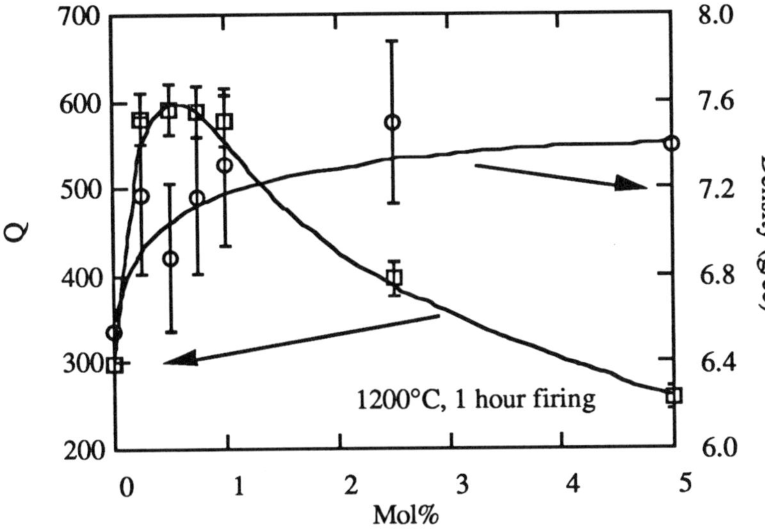

Figure 1 : Effect of Mn additions on the Q and density of PZC materials. The Q was measured at 450±25 MHz. (Error bar is ± 5% of the mean value).

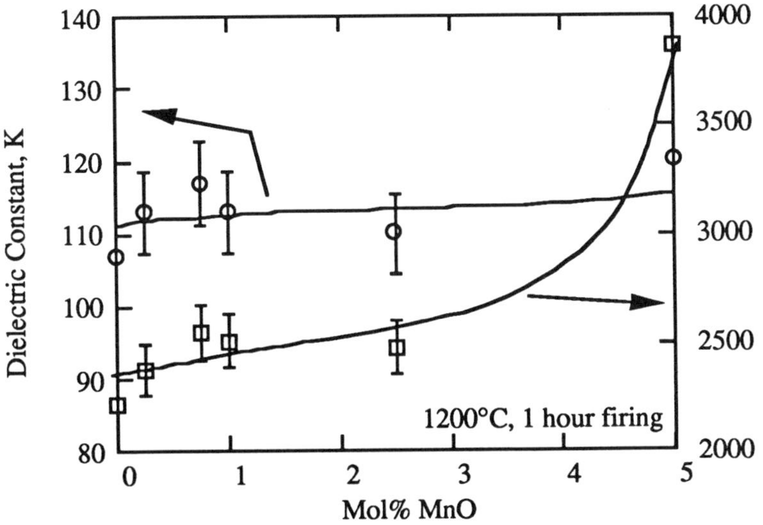

Figure 2 : Effect of Mn addition on the high frequency dielectric properties of PZC. (Error bar is ± 5% of the mean value).

Several samples with the same Mn contents were prepared with either MnO or MnO_2 to determine whether the two forms of manganese show different effect on the dielectric properties of PZC. No appreciable difference can be found between the materials prepared using two different oxides of manganese. This may be related to the fact that both oxides converted to Mn_3O_4 when heating in air to about 1000°C.[10] To see if Curie temperature of the sample was affected by Mn additions, the temperature dependence of K were measured for samples with 0.75 mol% and without MnO additions. Results indicate that the Curie temperature of these samples was about the same at 235°C. Phase analysis also shows no significant difference between the undoped and Mn-doped compositions within the detection limit of X-ray diffractometer and no second phase can be found in the samples.

(3) Effect of firing condition on PZC properties:

The sintering condition has also shown profound effect on the dielectric loss when MnO is added to the samples. For samples without MnO addition, the Q was seen (Figure 3) to increase with increasing firing temperatures up to 1280°C. A different trend was seen for MnO doped samples. Depending on the concentration levels, the Q may either increase or decrease with the sintering temperatures up to about 1250°C. Beyond that, higher sintering temperatures show an adverse effect on the Q for MnO modified samples. It is not clear at the moment about the exact cause of this behavior. Further studies will be required to clarify this point.

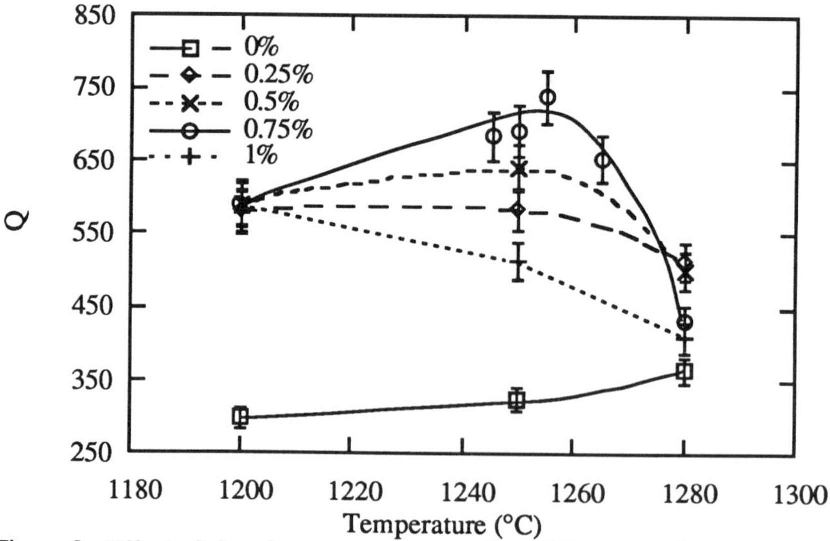

Figure 3 : Effect of sintering temperature on Q at different MnO contents.

From these results, it appears that in order to achieve high Q, both the MnO content and sintering conditions need to be controlled. Figure 4 shows the combined effect of MnO content and sintering temperatures on Q of PZC materials. The Q values are relatively insensitive to the amount of MnO in the samples when sintered at 1200°C. At 1250°C, the Q increases with MnO content levels up to 0.75 mol%, above that level, the Q decreases. As the sintering temperature is increased to 1280°C, a reverse effect is seen and samples with higher MnO content exhibit lower Q values. In this study, samples with 0.75 mol% MnO addition have shown a highest average Q value of 745 when sintered at 1250°C for 2 hours.

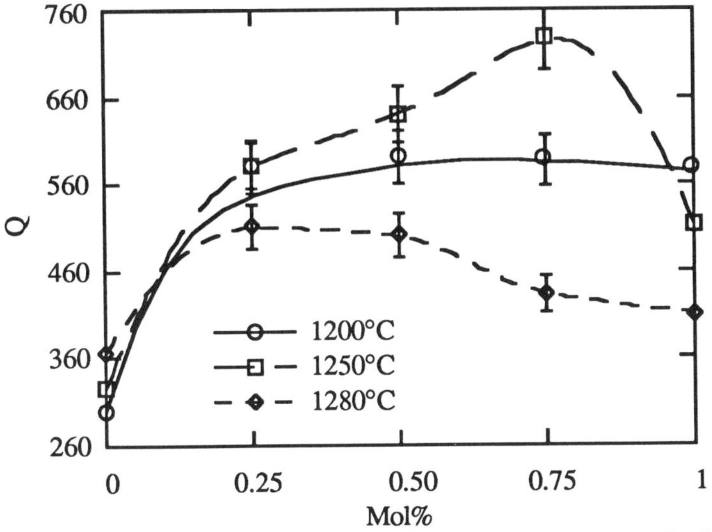

Figure 4: Effect of MnO dopants and firing temperature on Q of PZC.

(4) Possible mechanism for Q improvement

It is well known [11] that the porosity and grain boundaries in the ceramic have detrimental effects on dielectric Q. It is also known that for compounds having domains, the existence of domain walls can contribute to the dielectric loss due to domain wall vibration. Other dielectric loss mechanisms such as the electrical conduction of the dielectric, the time lag of the orientation of permanent dipoles, the accumulation of electric charges on the grain boundaries within the mixture, and the mechanical interaction between phases had also been proposed by Bunag and Koenig [12]. In order to find the main reason responsible for the enhancement of Q by Mn additions, several likely mechanisms were examined.

The samples were studied by the scanning electron microscopy to determine the effect of MnO addition on the microstructure of PZC ceramics. Results indicate that MnO is an effective sintering aid to reduce the porosity and promote grain growth in PZC, as compared with samples without Mn addition. However, as shown in Figure 1, samples with higher Q do not necessarily exhibit higher densities. This indicate that the reduction on porosity is not a dominating factor for Q improvement seen in Mn doped PZC. The effect of grain size of PZC materials on Q was reported previously [4] which showed that the Q is directly related to the grain size of the sample, with larger grain samples exhibiting higher Q. Larger grain size can reduce the grain boundary area which should help to reduce the dielectric loss.

Another possibility is related to the reduction of domain wall mobility inside PZC. As mentioned earlier, additives with lower valence cations have shown the effect to reduce the dielectric loss of PZC. Similar improvements on dielectric loss were also seen on Mn doped PLZT-PZN and PZT [13]. Originally, it was thought that the Q improvement seen in MnO doped PZC can be related to the reduction of domain wall vibration. However, if the Q improvement of MnO is caused by the reduction of domain wall vibration, then samples containing higher MnO should exhibit higher Q. But this is not to be the case as can be seen in Figure 1. Moreover, MnO doping in PZC is capable to raise the Q nearly 250% higher than the undoped one. It is a much more effective additive when compared to the 5% to 20% Q improvement made by other divalent cations shown in Table I. Other more important loss reducing mechanism other than the domain wall vibration reduction must have taken placed for the MnO doped samples. This observation is in agreement with the conclusion of Lanagan et al [6].

The addition of small amount of Mn was seen to improve the Q and sinterability of other high frequency dielectrics such as $Ba(Mg_{1/3}Ta_{2/3})O_3$, $Ba(Zn_{1/3}Ta_{2/3})O_3$ and $Ba_2Ti_9O_{20}$ [9,14]. Mn was considered to act as an oxidizing agent [14] in these compounds that reduce the oxygen vacancies located at the grain boundary region. It is possible that similar effect occurred to the MnO added PZC. The defect equation for oxygen vacancy can be shown as

$$O_O^x \;===\; V_O^{\bullet\bullet} + 2\,e' + 1/2\,O_2 \qquad (1)$$

where $V_O^{\bullet\bullet}$ is the oxygen vacancy and e' is the extra electron which can be a negatively charged cation. The existence of these defects can increase the conduction of the ceramic, hence reducing the Q. Although both oxygen vacancies and electrons can contribute to the loss, the loss is likely to be dominated by the electronic type defects. Since the charge mobility of ionic type defect like oxygen vacancy is much lower than the electronic type defects such as electron or electron hole, the dielectric loss at high frequencies should mainly coming from the more mobile electronic defects. Therefore, a possible scenario for Q improvement would be for MnO to act as an oxidizing agent that reduce the amount of oxygen vacancies. During this process, the electronic charge carriers are eliminated when oxygen vacancies are compensated as shown in the reverse reaction of equation (1). However, further study in the future is required to provide direct evidence and to pinpoint the exact mechanism responsible for Q improvement by manganese.

From the above discussion, it appears that MnO not only acts as a sintering aid to reduce the porosity and enhance the grain growth, it is also an effective additive to enhance the Q of PZC at high frequencies, possibly by the elimination of electronic type defects.

(5) Temperature compensated $PbZrO_3$-$(Sr,Ca)TiO_3$ Composite stripline:

Two microstrip resonators were formed with the Mn modified PZC and $(Sr,Ca)TiO_3$ (SCT) substrates. These substrates were prepared by conventional tape casting and sintering method. The microstrip lines were formed using electro-deposited thin film copper metallization to build up the thickness to 0.01 mm. A photolithography method was used to accurately define the line length. The Mn-PZC microstrip exhibited a T_f of - 1170 ppm/°C, a Q of 93 at 850 MHz with a K of 113 while the SCT sample showed a T_f of +583 ppm/°C, a Q of 150 at 388MHz with a K of 195. Notice that the difference in resonance frequencies are due to the different K values for the two materials. The two microstrips were sandwiched together by reflow soldering using a solder paste to form a composite stripline resonator. As shown in Figure 5, the resonator showed a T_f of - 68.7 ppm/°C from -25°C to 65°C with a Q of 136 at 764 MHz and a combined K of 133. These data indicate that the property of conventional sintered Mn doped $PbZrO_3$ may be used to construct temperature compensated composite stripline filters or resonators for high frequency applications.

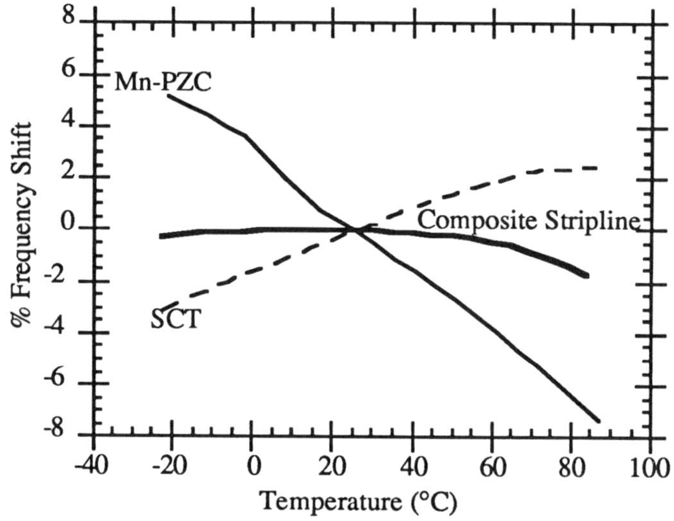

Figure 5: The percent frequency shift verses temperature curves for the $(Sr,Ca)TiO_3$, Mn-doped $PbZrO_3$, and a composite stripline made by these two materials.

CONCLUSION

The effect of oxide additives on the dielectric properties of lead zirconate were investigated with the aim to reduce the losses at high frequencies using conventional sintering method. The addition of small amounts of manganese was found to increase the dielectric Q significantly without reducing the dielectric constant. The Q factor was increased to 745 at 450 MHz by adding 0.75 mol% MnO and 2 mol% CeO_2 to lead zirconate when sintered at 1250°C for 2 hours. The reduction of dielectric loss is attributed to the reduction of electronic type defects as well as improvement in microstructure when using MnO as sintering aid. Other additives investigated were not effective in improving the Q without adversely affecting the K or vice-versa. The sintering condition has a dominant effect on the properties of Mn added lead zirconate. The lead zirconate ceramic with negative T_f and improved dielectric characteristics has been used to demonstrate a temperature compensated stripline with low T_f and high effective dielectric constant for high frequency applications.

REFERENCES

[1]. K.Wakino, "High Frequency Dielectrics and Their Applications", *Proc.IEEE International Symp. Appl. Ferroelectr.*, 97-106 (1986).

[2].A.E.Paladino, "Temperature Compensated $MgTi_2O_5$-TiO_2 Dielectrics", *J.Am.Ceram.Soc.,* **57**, 168-169 (1971).

[3]. K.Tatsuki, K.Murano, T.Kawamura, H.Kato, S.Yano, and S.Nishigaki, " Dielectric Properties of PbO-ZrO_2-REO_x Ceramics at Microwave Frequencies", *Jpn. J. Appl. Phys.*, **26**, Suppl.-2, 80-82(1987).

[4].K.Murano, K.Tatsuki, S.Nishigaki, S.Yano and H.Kato, "Microwave Dielectric Properties of Pb(Zr,Ce)O_3 Ceramics", *Proc. 3rd US-Japan Seminar on Piezoelectric And Dielectric Ceramic Materials, Toyama, Japan*, Paper **T-1**, 49-50 (1986).

[5].J.Kato, "Material Produces Small Resonators with High Dielectric Constant", *Jpn.Elect. Eng.*, 114-118 (Sept.1991)

[6].M.T.Lanagan, J.H.Kim, Sei-Joo Jang, and R.R.Newnham, "Microwave Dielectric Properties of Antiferroelectric Lead Zirconate", *J.Am.Ceram.Soc.*, **71**, 311-316 (1988).

[7]."Standard test method for effective series resistance (ESR) and capacitance of multilayer ceramic capacitors at high frequencies", *ASTM F752-82*, 729-739 (1982).

[8].B.Jaffe, W.R.Cook, and H.Jaffe, "*Piezoelectric Ceramics*", Academic Press, 154-159 (1971).

[9]. S.Nomura, " Ceramics for Microwave Dielectric Resonators", *Ferroelectr.*, **49**, 61-70 (1983).

[10]. R.D.W. Kemmitt, pp 798-805 in "*Comprehensive Inorganic Chemistry* " edited by J.C. Bailar Jr.,H.J.Emeleus, R. Nyholm, and A.F. Trotman-Dickenson, Pergamon Press, New York (1973).

[11]. J. M. Herbert, " *Ceramic Dielectrics and Capacitors* ", Gordon and Breach Science Publisher, pp 98 and 126 (1985).

[12]. M.M.Bunag, and J.H.Koenig, "Ultra Low Loss Ceramic Dielectrics ", *J.Am.Ceram.Soc.* **42,** 442-47(1959).

[13].S.Nomura, K.Kaneta, J.Kuwata, and K.Uchino, "Electromechanical Properties of Piezoelectric Ceramics in the PLZT-PZN System", *Jpn.J.Appl.Phys.*, **20**, 179-182, (1981, Suppl-4).

[14].S.Nomura, K.Tamaya, and K.Kaneta, "Effect of Mn Doping on the Dielectric Properties of Ba2Ti9O20 Ceramics at Microwave Frequency", *Jpn.J.Appl.Phys.*, **22**, 1125-28(1983).

PHOTOIMAGEABLE THICK FILM METALLIZATION ON MICROWAVE SUBSTRATE FOR WIRELESS APPLICATIONS

D.D. Chang and H.C. Ling
AT&T Bell Laboratories
Engineering Research Center
P.O. Box 900, Princeton, NJ

ABSTRACT

The emerging personal wireless communication products require miniaturized radio frequency (RF) components and substrates. For some applications, one proposed approach is to use high dielectric constant (k) ceramic substrates to shrink the length of RF circuits and to directly fabricate certain RF components (filters, capacitors) onto the substrate using distributed transmission line structures. This approach depends on the capability of metallization process in defining the narrow conductor line width of the transmission lines.

We have evaluated the capability of photoimageable thick film process, developed by Du Pont, in defining fine gold conductor lines. A variation of ±0.1 mil was measured on transmission line widths between 1 and 10 mils, fabricated on $Ba_2Ti_9O_{20}$ ceramic substrates (k=38) with a surface roughness of 40 microinches (1 μm). This translates into a maximum variation of ±10%. The measured S_{21} value at 2 GHz is in good agreement with the calculated S_{21} value from dielectric and conductor losses. Using an interdigital bandpass filter design, the electrical characteristics of filters fabricated by the photoimageable thick film process was identical to similar filters made by thin film metallization process. Thus the thick film process may represent a low cost, high volume manufacturing option.

1. INTRODUCTION

The emerging personal wireless communication products require miniaturized radio frequency (RF) components and substrates. The miniaturization comes from the shrinkage of discrete component's physical dimensions as well as fabricating finer line width and space on the substrate. Fine line process capability is needed, because it determines how well the control of RF transmission line impedance and associated energy transmitting efficiency.

The specified impedance value is governed by transmission line equations; the properties and physical dimensions of the transmission lines and the surrounding dielectric material determine the impedance. For a target line impedance and a given substrate thickness, a substrate with a higher dielectric constant requires the finer line width. For tight impedance control, it is necessary to control the conductor line width to within ±5 to 10%. This corresponds to ±0.05 to 0.1 mil on a 1 mil width line.

Under a quasi-TEM mode propagation, the characteristic impedance of a microstrip line is described by EQ (1). For a narrow strip line (w/h < 3.3),

$$Z_o = \frac{119.9}{\sqrt{2(k+1)}}\left[\ln\left\{\frac{4h}{w}+\sqrt{16\left(\frac{h}{w}\right)^2+2}\right\}\right] \qquad (1a)$$

For a wider strip line (w/h > 3.3),

$$Z_o = \frac{119.9\pi}{2\sqrt{k}}\left[\frac{w}{2h}+\frac{\ln 4}{\pi}+\frac{\ln(e\pi^2/16)}{2\pi}\left(\frac{k-1}{k^2}\right)+\frac{k+1}{2\pi k}\left\{\ln\left(\frac{\pi e}{2}\right)+\ln\left(\frac{w}{2h}+0.94\right)\right\}\right]^{-1} \qquad (1b)$$

where e is the exponential base (e=2.718), w is the conductor width, h is the dielectric thickness and k is the dielectric constant of the substrate. The effect on conductor width reduction due to the increase of substrate dielectric constant, for a target line impedance of 50 Ω, is shown in Fig. 1.

Figure 1 shows high-k ceramics can bring RF components horizontally closer together using narrow-width conductors. Along the signal propagation direction, the signal velocity (group velocity) is inversely proportional to the square root of the effective dielectric constant, $\sqrt{k_{eff}}$,and this velocity determines the quarter-wavelength. Thus a higher dielectric constant substrate results in shorter quarter wavelength on the transmission line. The quarter wavelength is an important parameter in RF circuits, such as impedance transformer and matching network. A high k substrate reduces the length of these circuits. In addition, some discrete RF components (such as filters and capacitors) can be directly fabricated onto the substrate using distributed transmission line structures, achieving circuit height as well as length reduction.

The advantages of utilizing high k ceramic substrates in miniaturizing RF circuits and building planar RF components (filters, couplers) depends on the ability of the metallization process in defining the narrow conductor line width and length. Thin film metallization technique has been used in realizing such circuits. Thick film metallization process may represent an alternate, less expensive manufacturing option if similar fine line definition can be achieved. In a previous report [1] we have shown that it is difficult to achieve good resolution on fine lines (4 mil or less) using conventional thick film paste and screen-printing techniques. Recently, photoimageable thick film paste which is capable of producing conductor line width down to 1 mil was developed by Du Pont[2]. We have evaluated this new thick film process by fabricating microstrip transmission lines and band-pass filters on high-k ceramic substrates. This report describes the electrical characteristics of these circuits at RF frequency from 50 MHz to 10 GHz.

2. TEST CIRCUITS

Figure 2 shows the physical layout of the test circuit. Each circuit substrate contains 18 microstrip transmission lines and 8 interdigital bandpass filters. The design of the transmission lines was targeted at 1, 4 and 9.8-mil line widths; and the filter's operating center frequency was designed at 2 GHz.

2.1 Microstrip Transmission Line

The intend of fabricating microstrip transmission lines was to evaluate their line width variations which can be judged by impedance measurements using time-domain reflectometry technique and by scanning electron microscopy (SEM). It also allows us to evaluate the conductors' losses based on the transmission coefficient, S_{21}, measurements using a network analyzer.

Using a 27-mil thick ceramic substrate with a k value of 38, the theoretical impedance of these lines can be calculated using Equation 1. The S_{21} response relating to the conductor loss (α_c) and dielectric loss (α_d) can be calculated[3].

2.2 Interdigital Bandpass Filter

Bandpass filters are the most essential components in today's wireless communication equipment. As discrete components, these filters are made from ceramic coaxial block based on the combline filter design which consists of a series of parallel coupled quarter wavelength resonators. For a given operating center frequency, the resonator's length (ie. filter's size) is proportional to $1/\sqrt{k_{eff}}$. To miniaturize the filter's size, a high-k material is desirable. However, when the application frequency increases (to 2 GHz or above) and the quarter wavelength decreases, a low-k material is frequently used instead to avoid having to handle and fabricate small parts to high precision. Therefore the challenge of making miniaturized ceramic block filters, using high-k material at 2 GHz, comes from cost effective manufacturing of each resonator at a specified quarter-wavelength length (ie. 8-mm, if k=38 is used).

Integrated planar bandpass filters on a substrate can be realized, using the same high-k material for block filters, by distributed transmission line structure at lower cost. While the loaded-Q of the filter will be reduced from the intrinsic Q of the material due to the presence of the microstrip's air/substrate interface, it is probably not important for use in many low power, portable wireless terminals.

Microstrip tapped-line interdigital bandpass filters have been considered by many designers due to their space and cost-saving advantages over conventional ceramic block filters. Design details of these filters has been published [4]. Since our focus was to evaluate the new thick film metallization process and compare that to a conventional thin film process using a simple filter circuitry, therefore we did not optimize the circuit's physical layout to achieve the best electrical performance.

The equivalent circuit and physical layout of the filter is shown in Figure 3 [5]. It is a 3-pole bandpass filter with a center frequency at 2 GHz. The physical layout consists of three parallel coupled resonators with ground via at the end of each resonator. Since the design requires grounding one end of the resonators, vias were drilled to connect to ground plane on the backside of the substrate. The ground vias have the effect of providing series inductance with the microstrip lines, therefore determining the bandpass filter's center frequency.

3. PHOTOIMAGEABLE THICK FILM PROCESS

State-of-the-art thick film processing that rely solely on screen-printing can probably achieve reasonable line definition for lines wider than 5 mils (125 μm). To achieve line width control between 4 (100 μm) and 1 mil (25 μm), we used a photoimageable thick film gold (Au) paste originally developed by Du Pont for MCM applications.[6] [7]. We provided the ceramic substrates and mask artwork on a mylar sheet and Du Pont Electronics performed the metallization process to transfer both standard transmission lines and filters onto the ceramic substrates. While a variety of substrates with k from 9.8 (alumina) to 80 ((Ba,Sm)Ti oxide) were prepared, we will only discuss results from the AT&T dielectric[8] , which has a k value of 38 and a thickness of 27 mil.

The metallization process involved print/dry/expose/develop/fire. The printing was done at a single pass at 25.4 cm/sec. using a 400 mesh stainless steel screen. This covered the entire substrate with the paste. The substrates were allowed to sit at room temperature for 10-15 minutes (leveling), followed by drying at 80°C for 15-25 minutes. The exposure was done under a UV Hg/Xe light source at 120-400 mJ/cm^2, corresponding to 10-40 sec. with a 1 KW lamp. The developing process was conducted in a conveyorized, spraying unit with aqueous 0.8-1.0% sodium carbonate at 30°C. The samples were rinsed with water immediately after development and any excess water was removed by blow-drying with air. The gold conductor layer was fired in a belt furnace with a 60 minute cycle and 10 minutes at the peak temperature of 850°C. There was a 50% shrinkage normal to the printed surface. With a printed and dried thickness of 14-16 μm, the fired conductor line thickness was 7-8 μm. In the plane of the printed surface, there was a nominal shrinkage of 0.3 mil upon firing, which was built into the transmission line design. The back-side conductor was Du Pont's standard 6160 thick film silver conductor paste, printed and fired on the substrates after the application of Fodel® 5956.

Figure 4 shows the metallized transmission lines and a number of interdigital filters on a variety of ceramic substrates, typically 2 in.(5.18 cm) by 2 in.(5.18 cm). Figure 5 shows the higher resolution photomicrographs of the transmission lines and an interdigital filter. While the design line widths were 1, 4 and 9.8 mils, the measured line widths were 1.1, 4.1 and 9.8 mils, respectively. Figure 6 shows the SEM micrographs of the transmission lines on k=38 substrate. The surface roughness of these substrates is about 40 micro-inches (1 μm)[9] compared to the typical 8 micro-inches (0.2 μm) seen in alumina substrates. The line width variation is about ± 0.1 mil (2.5 μm), which amounts to ±10% for the 1.1 mil line and less than ± 5% for the wider lines. At this limit, the variation will be affected by the smoothness of the substrate surface. It is anticipated on a alumina substrate, the line width variation will be less than ±5% for the 1 mil line. This is much better than the ± 1 mil (25 μm) variation obtained using the Micro-pen writing technique reported previously[1].

4. RESULTS AND DISCUSSIONS

4.1 Microstrip Transmission Characteristics

The S_{21} response between 50 MHz and 5 GHz for the three transmission lines (length = 1.27 cm and width = 1.1, 4.1 and 9.8 mils) are shown in Figure 7. The average downward slope of the curves represents the loss along the length of the transmission lines as a function of frequency. The long wavelength undulation is probably related to impedance mismatch, since it does not appear in the S_{21} curve of the 4.1 mil line, which was designed to have an impedance of 50 Ω. Table I shows the calculated conductor loss and dielectric loss of the transmission lines and the corresponding S_{21} values are compared to the measured S_{21} values at 2 GHz. They are in good agreement to within 0.4 dB.

Table I. Transmission Lines Loss Comparison

width	Z	α_d	α_c	α	$S_{21}(\alpha)$	S_{21}(meas)
(mil)	(Ω)	(dB/cm)	(dB/cm)	(dB/cm)	dB	dB
1.1	68.2	0.00082	0.1476	0.1484	0.819	0.48
4.1	50.9	0.00084	0.0845	0.0853	0.471	0.24
9.8	38.9	0.00086	0.0535	0.0544	0.300	0.49

4.2 Interdigital Bandpass Filters

The test filters' frequency response were characterized by their forward transmission coefficient, S_{21}, measured from 50 MHz to 10 GHz. Figure 8 compares the S_{21} response on test filters made by a thin-film process versus the photoimageable thick film process on the same substrate material. Basically, the filters made from the thick film process performed the same as the ones made by the thin film process, within the experimental error. Since three ground vias are needed to achieve the 2 GHz design center frequency, during the via drilling process, we realized that consistency in the forming of the ground vias determines the reproducibility of the filters' electrical performance and tolerance. In a manufacturing environment, laser drilling of the vias will probably be needed to maintain quality control.

5. CONCLUSIONS

We have evaluated the capability of photoimageable thick film process, developed by Du Pont, in defining fine gold conductor lines. A variation of ±0.1 mil was measured on transmission line widths between 1 and 10 mils, fabricated on $Ba_2Ti_9O_{20}$ ceramic substrates (k=38) with a surface roughness of 40 microinches. This translates into a maximum variation of ±10%. The measured S_{21} value at 2 GHz is in good agreement with the calculated S_{21} value from dielectric and conductor losses. Using an interdigital bandpass filter design, the electrical characteristics of filters fabricated by the photoimageable thick film process were identical to similar filters made by thin film metallization process. Thus the thick film process may represent a low cost, high volume manufacturing option.

6. ACKNOWLEDGEMENT

We wish to thank H.M. O'Bryan for providing the $Ba_2Ti_9O_{20}$ substrates used in this study, Terry Suess of Du Pont Electronics for performing the metallization process and Steve Barthelmes, Mike Zierdt and Mike Evans for providing the thin film filters and helpful design discussions. We also acknowledge R. Woods for performing the SEM microstructure analysis.

REFERENCES

1. H.C. Ling, H.M. O'Bryan and J. Thomson, Jr., "Thick film Metallization on Microwave Substrates for Wireless Communication," presented in Ceramics for Wireless Communication Symposium, 95th American Ceramic Society annual meeting, Cincinnato, OH, 4/93; and the Sixth US-Japan Seminar on Dielectric and Piezoelectric Ceramics, pp. 227-230, 11/93.
2. Fodel® 5956 gold conductor paste, Du Pont Electronics, Research Triangle Park, NC.
3. R.A. Pucel, D.J. Masse & C.P. Hartwig, "Losses in Microstrip," IEEE Transaction on MTT, vol. MTT-16, no. 6, 1968.
4. J.S. Wong, "Microstrip tapped-line filter design," IEEE Trans. on Microwave Theory and Technology, vol. MTT-27, no. 1, Jan, 1979, pp.44-50.
5. E.G. Cristal, "Tapped-line coupled transmission lines with applications to interdigital and combline filters," IEEE Trans. on Microwave Theory and Technology, vol. MTT-23, no. 12, Dec., 1975, pp. 1007-1012.
6. Fodel® 5956 gold conductor paste, Du Pont Electronics, Research Triangle Park, NC.
7. Jean-Pierre Cazenave, "Photoimageable thick film materials: an alternative to advanced MCM's technologies," Proceedings of Int. Conf. & Exhibition on Multichip Modules, pp.301-306, Denver, 4/93.
8. H.M. O'Bryan, J. Thomson, Jr. and J.K. Plourde, "A new BaO-TiO2 compound with temperature-stable high permittivity and low microwave loss," J. Am. Ceram. Soc., vol. 57, p.450 (1974).
9. measured by Dektak IIA, Vecco-Sloan, Santa Barbara, CA

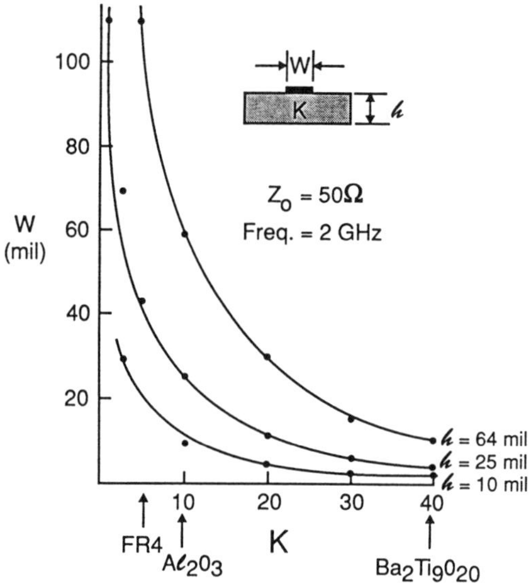

Figure 1. Effect on conductor width (w) reduction as a function of substrate dielectric constant (k).

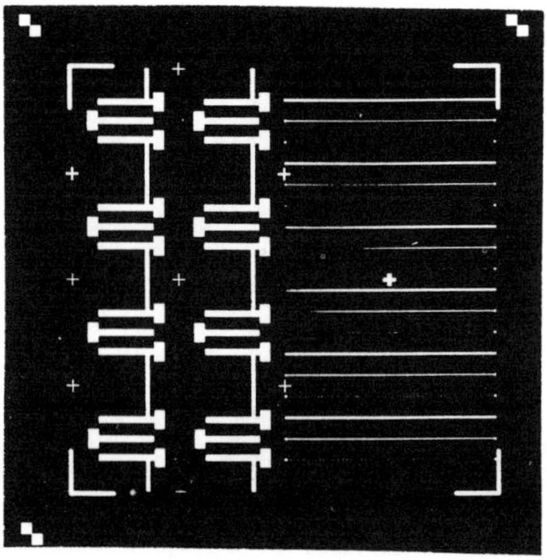

Figure 2. Physical layout of test circuits.

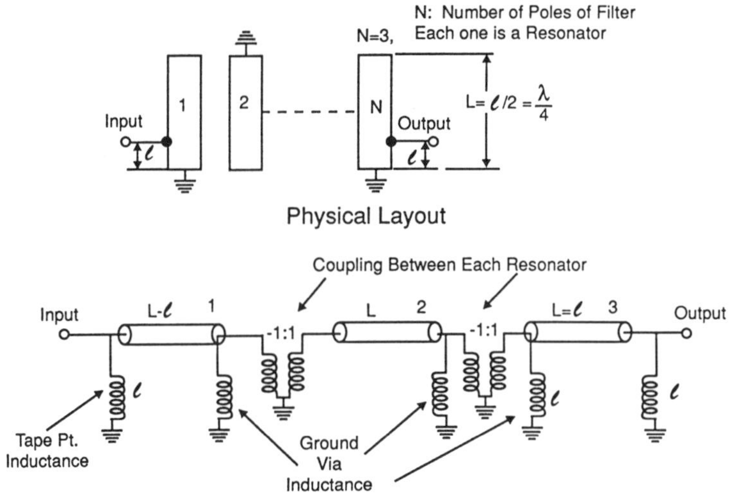

Figure 3. Equivalent circuit and physical layout of interdigital bandpass filter.

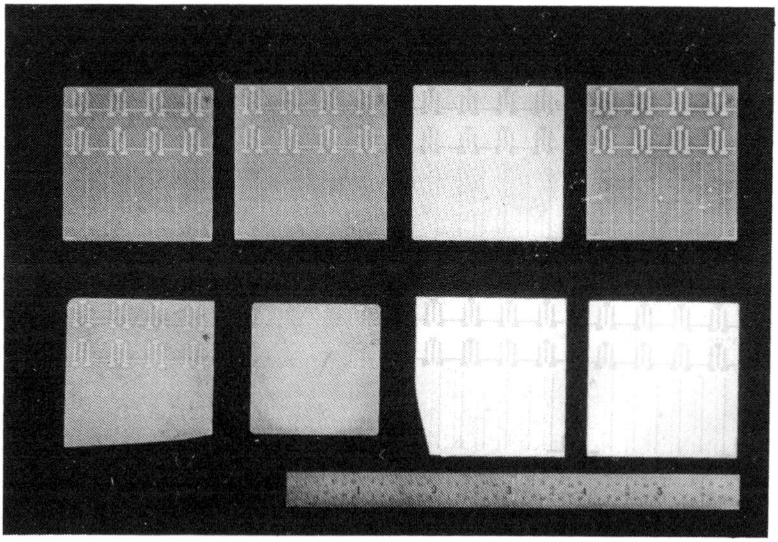

Figure 4. Metallized transmission lines and interdigital filters on a variety of ceramic substrates.

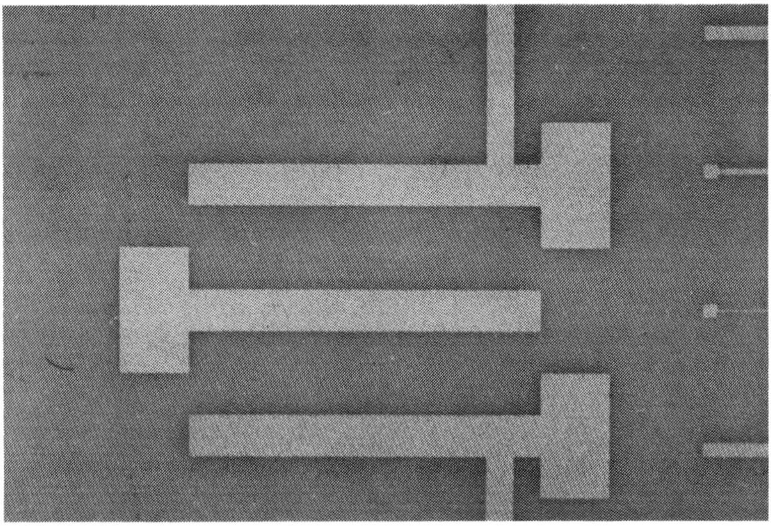

Figure 5. Optical micrograph of transmission lines and an interdigital filter on a $Ba_2Ti_9O_{20}$ ceramic substrate.

Figure 6. SEM micrographs of the transmission lines with widths of 9.8, 4.1 and 1.1 mil and a thickness of 7-8 µm.

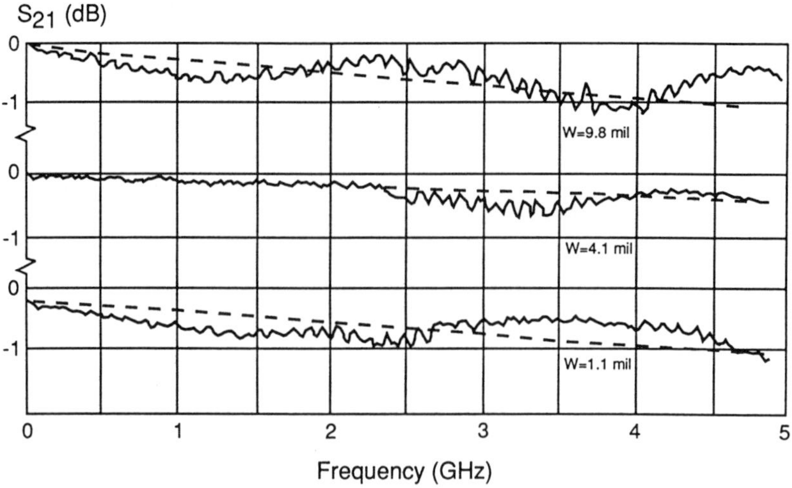

Figure 7. S_{21} of transmission lines as a function of frequency between 50 MHz and 5 GHz.

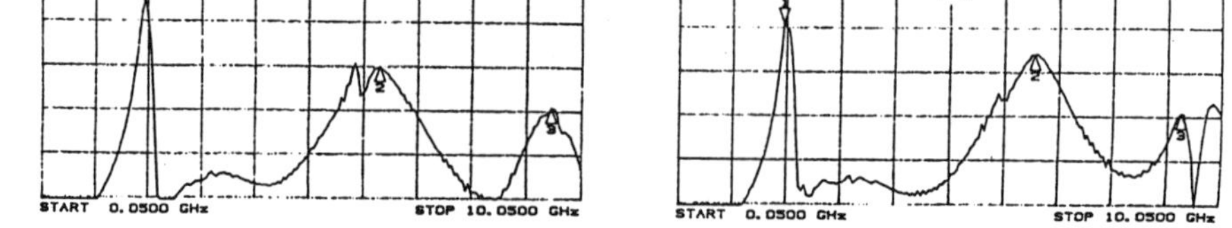

Figure 8. S_{21} of interdigital filter as a function of frequency between 50 MHz and 10 GHz fabricated by (a) thin film process, versus (b) the photoimageable thick film process.

DENSIFICATION BEHAVIOR OF METAL AND DIELECTRIC FILMS ON RIGID SUBSTRATES

Guo-Quan Lu, Jaecheol Bang, JoonWon Choe, and Jesus N. Calata
Virginia Polytechnic Institute and State University
Blacksburg, Virginia 24061

ABSTRACT

We measured the densification kinetics and observed microstructures of constrained gold and glass+ceramic films on rigid substrates as well as those of the free films. In the gold which densifies via solid state diffusion, we found significantly retarded densification kinetics in the constrained films; the measured activation energies for densification suggest that the sintering mechanism in the free films was controlled by surface diffusion, and the mechanism was changed to bulk diffusion in the constrained films. In the dielectric glass+ceramic system, we also observed retarded densification kinetics in the constrained films but to a lesser extent than those found in the gold; activation energies for densification in the dielectric which densifies via a viscous flow mechanism were found to be the same in the constrained and free films.

INTRODUCTION

Constrained-film sintering is an important process in the manufacture of interconnecting packages for electronic and optoelectronic devices. For example, to fabricate hybrid electronic packages, porous metal ink films and inorganic dielectric green layers are constructed on rigid substrates and then sintered together at high temperatures[1]. The metal/dielectric structure forms the electrical interconnecting network of the package. During the sintering process, the porous materials are constrained from shrinking in the substrate plane and densified mainly by shrinking normal to the substrate. Constrained-film sintering can lead to defective packages. Common defects in hybrid packages are[1] high porosity in sintered films, and delamination or cracking at film/substrate interface. In multilayer cofired ceramic packages, part distortion or camber and via defects[2] are some major concerns to manufacturers. Packaging engineers currently are limited to trial and error solutions to minimize the defects, resulting in long product development cycles; a basic understanding on the kinetics of constrained-film sintering will enable them to develop systematic solutions to problems from constrained sintering.

Recently we developed[3-5] optical techniques capable of accurate measurements of densification profiles of free and constrained metal ink films and ceramic green tapes. We had used the techniques to understand defect developments in the cofiring process of a low-temperature cofired ceramic (LTCC) substrate[4]. In this paper, we report our results on the constrained-film sintering in a gold circuit paste and a borosilicate glass (BSG)+silica tape material; both of which could be used in a LTCC system.

EXPERIMENTAL

A. Sample Preparation

Gold Films

A DuPont gold circuit paste designed for low temperature cofired ceramic packages was used in this study. The paste had about 75% by weight of pure gold particles mixed in an organic binder vehicle. The gold particles were nearly spherical with an average radius of 2 μm. Ink films about 63 ± 2 μm thick were made by multiple screen-printing on a 0.5 mm thick alumina substrate partially diced into 6 mm x 6 mm squares. While still on the substrate, the films were subjected to a binder burn-out procedure by heating up to 450°C for two hours. There was good adhesion between the metal film and alumina after the burn-out. With the binder removed, a portion of the square film sample was carefully peeled off forming a free standing sample, and the remained portion formed a constrained sample. This procedure ensured the same initial microstructure in both the free and constrained films. The initial relative green density of the films was found to be 0.42 ± 0.01 determined by weighing a film and measuring its volume.

Sintering runs were done isothermally in a hot stage at temperatures between 650°C and 900°C. The ramp-up rate from 450°C binder burn-out to a sintering temperature was typically 100°C/min. In the constrained samples, while there was good adhesion between the gold film and alumina substrate, we expected and found no chemical reaction between the two materials after sintering.

BSG+Silica Films

A borosilicate glass (BSG) powder with an average particle size of 2 μm and softening point of 710°C was purchased from Sem-Com and a 99.5% pure silica powder having an average particle size of 2 μm was bought from Johnson Matthey. These two powders were mixed to give a silica content of 20% by volume. The powder mixture was ball-milled in an organic binder solution for 24 hours producing an uniform slurry. The premixed binder solution was donated by Palomar-MSI; it contained polyvinyl butyral (PVB) polymer binder, a plasticizer, a surfactant, and an ethanol+toluene solvent mixture. Samples were prepared by casting the slurry using a hand-held doctor-blade. Free films were formed by casting the slurry on silicone-coated mylar sheets for easy peel-off; constrained films were made by casting the slurry on 0.5 mm thick silicon substrates. The free films had an average thickness of 0.6 mm and the constrained films were approximately 0.2 mm thick, and both had the same initial relative density of 60%.

All the ceramic films were subjected to binder removal by heating up to 500°C for 2.5 hours; no densification was noted in both the free and constrained films after binder burn-out. Isothermal sintering was performed at temperatures between 715°C and 775°C. A three-minute rise time was allowed for ramp-up from the 500°C binder burn-out step. Like the constrained gold samples, there was good adhesion but no chemical reaction between the ceramic film and silicon substrate in the constrained ceramic samples.

B. Optical Setup

The densification kinetics of free and constrained films were measured using an optical setup. The setup can be arranged into two different configurations with minor adjustments of some optical components. One configuration is for determining width shrinkage profiles of free films, and the other for determining thickness shrinkage profiles of constrained films. Both of these had been described in detail in our previous papers[3,5].

RESULTS AND DISCUSSION

Gold Films

Figure 1 is a plot of the densification profiles (relative density vs. time curves) obtained on a free and constrained gold film from a sintering run at 650°C. These curves were calculated from the shrinkage profiles measured with the optical setup according to:

for free film : $\overline{\rho}(t) = \overline{\rho}_0 \cdot [w_0 / w(t)]^3$;

for constrained film : $\overline{\rho}(t) = \overline{\rho}_0 \cdot [d_0 / d(t)]$;

where $\overline{\rho}_0$ is the initial relative density (normalized to the final density of 19.3 g/cm^3); w_0 is the initial width of the free film and w(t) is the width after a sintering time t; d_0 is the initial thickness of the constrained film and d(t) is the thickness after a time t. The free film was assumed to shrink isotropically, a reasonable assumption for a powder compact with nearly spherical particles.

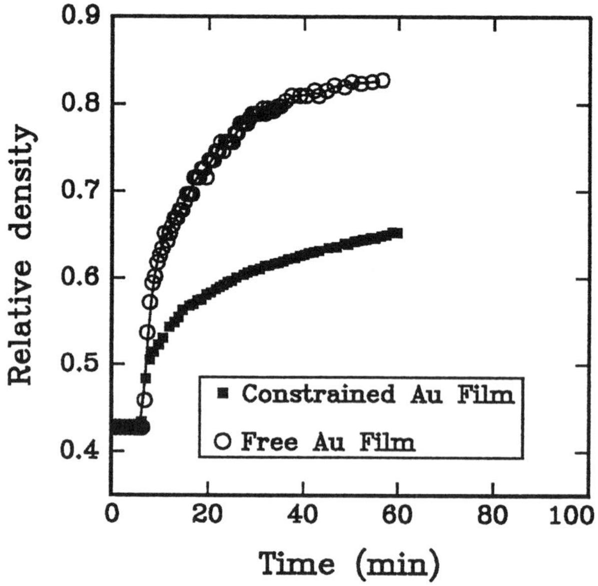

Figure 1. The relative density profiles of a constrained and freestanding DuPont film at 650°C.

The two curves clearly show that the densification rates of the constrained film were substantially lower than those of the free film; the initial rate was more than a factor of ten lower in the constrained film. Similar results were obtained on the two types of films sintered at other temperatures. Figure 2 is a plot of three pairs at three different temperatures, 650°C, 750°C, and 850°C. Note that only the portion of a curve at the sintering temperature was plotted. To show the extent of rate-retardation, we point out that the final density in the constrained gold film sintered at 850°C is close to that in a free film sintered only at 650°C, 200°C lower.

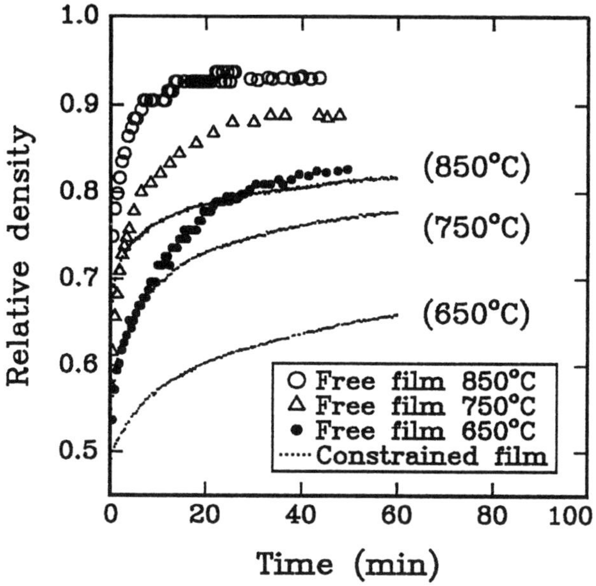

Figure 2. The relative density profiles of a constrained and freestanding DuPont film at 650°C, 750°C and 850°C.

Figure 3(a) and (b) are SEM micrographs taken in two pairs of films after two and half hours sintering at 650°C and 850°C, respectively. It is evident that more pores were left in the constrained films. A careful examination of the grains in Figure 3(b) also indicates that the grain size was nearly identical in both the free and constrained films. The same indication was also found in samples sintered at other temperatures, although the grain structures in the two micrographs shown in Figure 3(a) are difficult to see. These results suggest that the substrate constraint had no effect on the grain growth kinetics of the films although it had a drastic effect on the sintering kinetics.

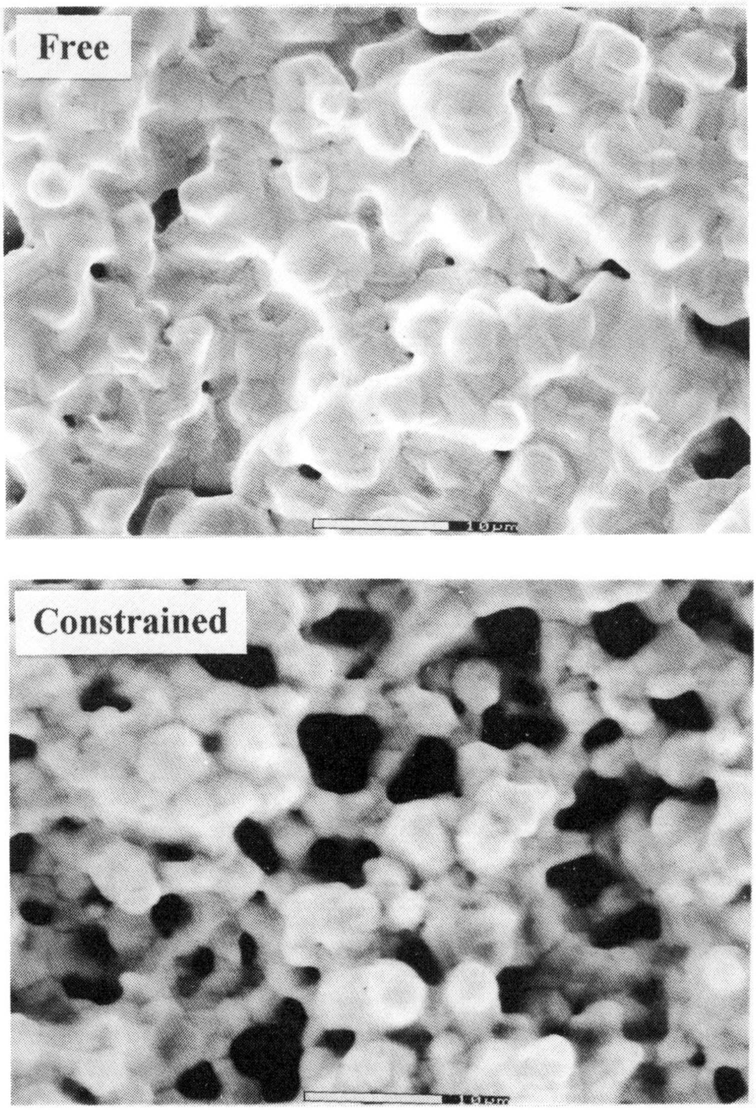

Figure 3(a). SEM micrographs of freestanding and constrained gold films sintered at 650°C

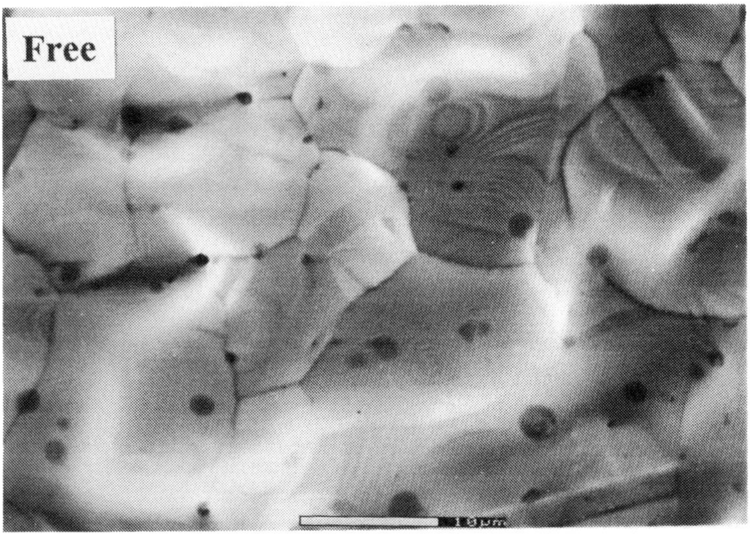

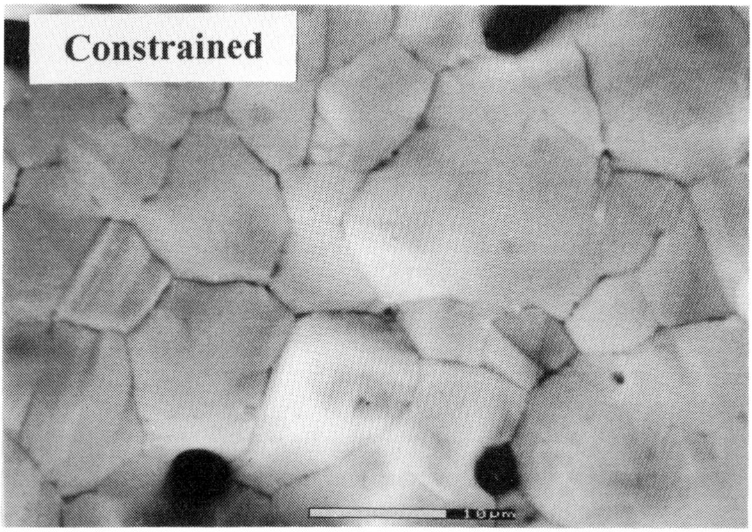

Figure 3(b). SEM micrographs of freestanding and constrained gold films sintered at 850°C

We further investigated the effect of substrate constraint on the energetic of the constrained-film sintering: the activation energy. A general equation for the densification rate of a porous material via solid state diffusion can be written as[6]

$$\dot{\bar{\rho}} = \frac{A}{d^n} \cdot \frac{\gamma V^{2/3}}{T} \cdot \exp\left(-\frac{Q}{RT}\right) \cdot f(\bar{\rho}) \tag{1}$$

where $\bar{\rho}$ is the relative density; d is the grain size; V is the molar volume; γ is the surface energy; T is the absolute temperature; Q is the activation energy; R is the gas constant; A is a constant independent of the other variables; and $f(\bar{\rho})$ depends only on $\bar{\rho}$. The activation energy Q is determined by plotting ($\dot{\bar{\rho}}T$) vs. (1/T) for a given relative density; the data can be extracted from measured densification profiles in Figure 2. Figure 4 is the activation energy plot for the constrained films at relative densities ranging from 0.63 to 0.70 and for the free films at 0.7 and 0.75. The graph gave fairly constant activation energy values for both the constrained and free gold films; the activation energy for sintering of free standing gold films was 90.1 ± 4.3 kJ/mole, and that of constrained films was 188.8 ± 6.7 kJ/mole. These values, respectively, are close to the activation energies for gold grain boundary diffusion (84.5 kJ/mole) and lattice diffusion (174.5 kJ/mole)[7].

Based on the above results, we suggest that the observed rate-retardation in the constrained films was due to a change in the dominant diffusion mechanism from grain boundary diffusion in the free films to lattice diffusion in the constrained films. The possibility of this mechanism change in constrained-film sintering had been predicted by Bordia and Scherer[8]. Figures 5(a) and (b) are illustrations, adapted from Bordia and Scherer, of the mass flow paths for densification in free and constrained films, respectively. In the free film, atoms flow along the grain boundaries into the neck regions, such as points A and B, causing the pores to shrink. In other words, densification is achieved by the transport of vacancies out of the pores along the grain boundaries to all the film surfaces. Here, the rate-limiting process is the atomic or vacancy diffusion along the grain boundaries that is responsible for bringing closer together the particles along all directions. On the other hand, in the constrained film as illustrated in Figure 5(b), the particles are physically locked from coming together in the substrate plane. Thus, the flow of vacancies through the grain boundaries in the directions along the substrate plane is restricted. Vacancies can only flow out in the direction normal to the substrate. That requires, as Bordia and Scherer pointed out, the diffusion of atoms through the bulk into pore regions and makes lattice diffusion the rate-limiting step for densification.

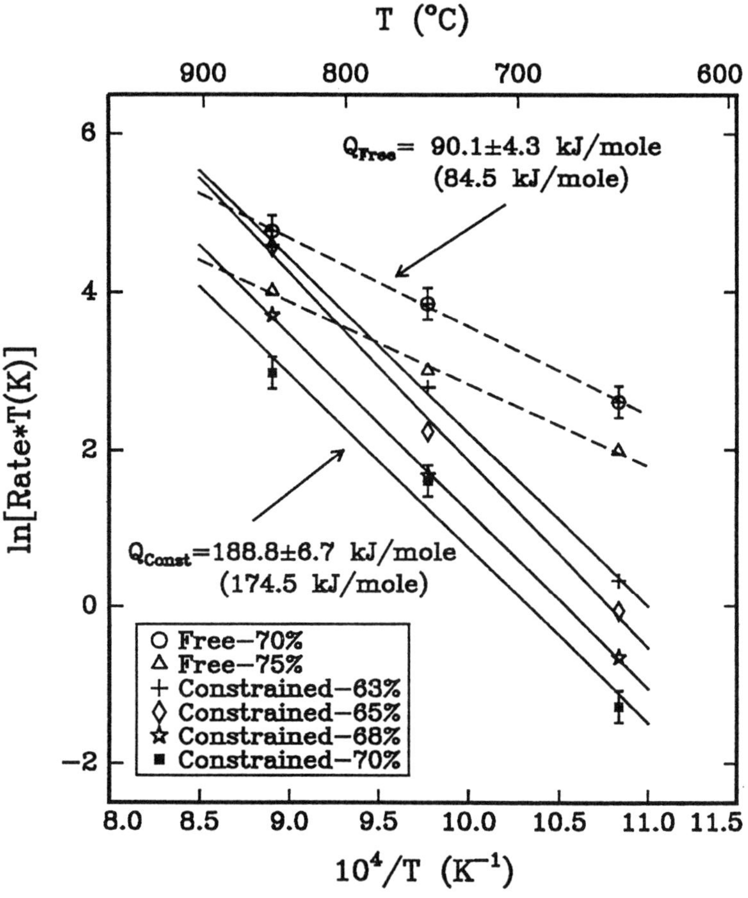

Figure 4. Activation energy plots for free and constrained DuPont gold films.

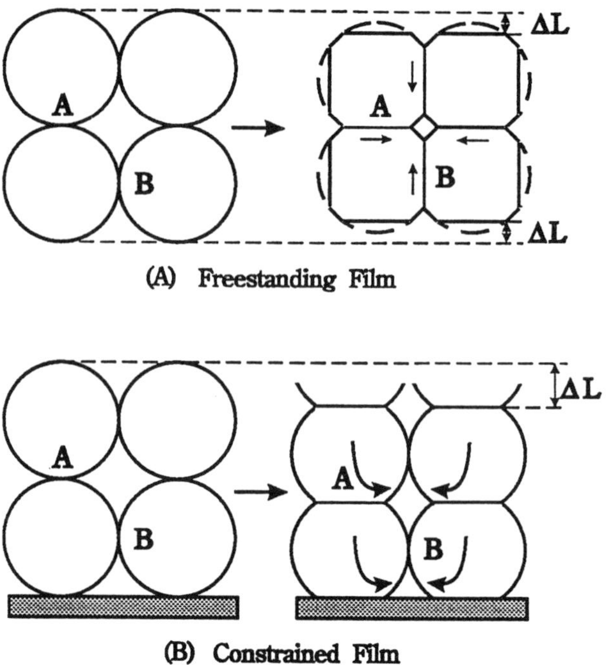

Figure 5. Schematic diagrams for the mass flow in free (A) and constrained (B) films.

<u>BSG+Silica Films</u>

Plotted in Figure 6(a) are the densification profiles of the free and constrained BSG+silica films sintered at 715°C, and in (b) at 775°C. As in the constrained gold, the densification rates were reduced in the constrained dielectric films compared with those in the free films. But, the rate-retardation was less severe in the dielectric than that in the gold. Shown in Figure 7 are SEM micrographs taken in the films after a three-hour isothermal sintering. These microstructures confirm the measured densification data, i.e. the constrained film is more porous than the free film at a given sintering temperature, and the films sintered at the lower temperature are more porous than those sintered at the higher temperature. Similar results were obtained in samples that were sintered at temperatures between 715°C and 775°C.

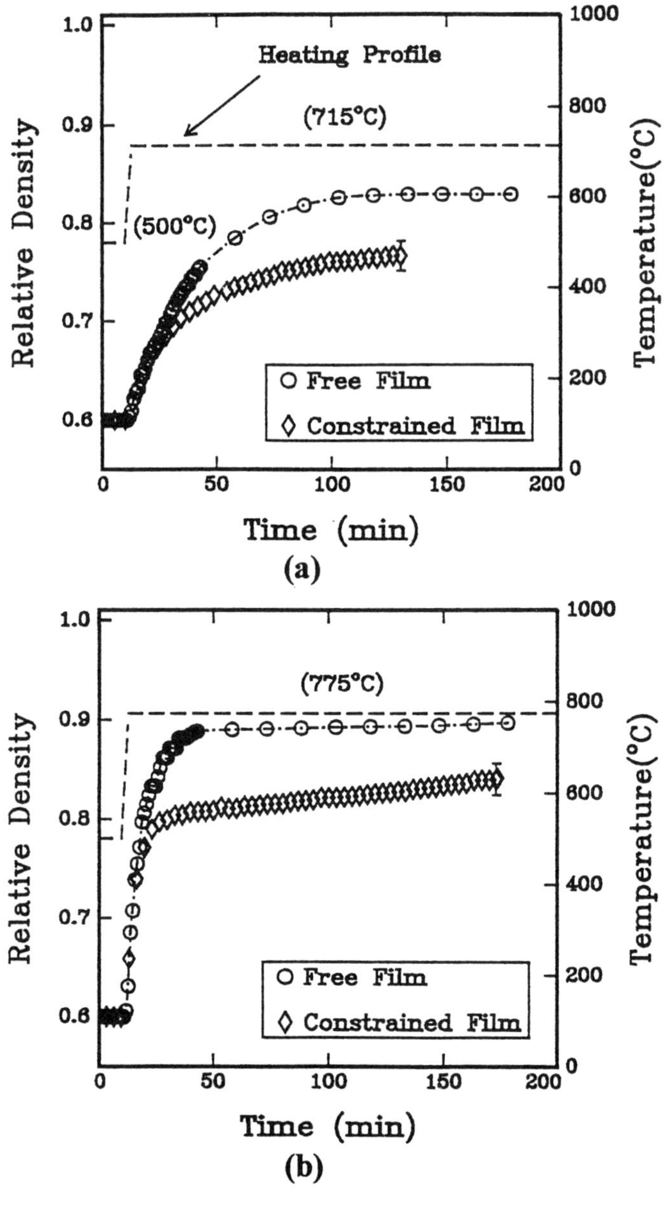

Figure 6. Densification profiles for free and constrained BSG+silica films sintered at (a) 715°C and (b) 775°C.

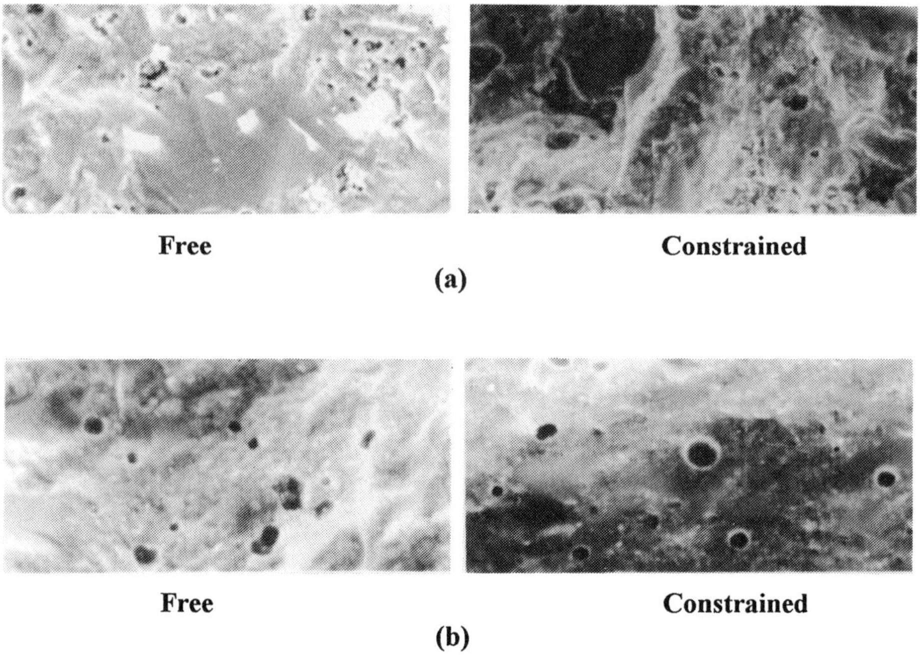

Figure 7. SEM micrographs of free and constrained glass films sintered at (a) 715°C and (b) 775°C.

Since the sintering runs were carried out at temperatures above the BSG softening point of 710°C, the predominant densification mechanism in the BSG+silica films is controlled by viscous flow of the BSG constituent. To determine the activation energy of densification, we adopt the generalized equation developed by Scherer[9] for the densification rate($\dot{\bar{\rho}}$):

$$\dot{\bar{\rho}} = (\mathrm{DF}) \times (\mathrm{KF}) \tag{2}$$

where DF is the driving force that depends on surface tension, initial pore size of the specimen, and a geometrical factor; and KF is the kinetic term inversely proportional to glass viscosity, η. η is expressed in terms of the activation energy (Q) and temperature (T) given below:

$$\eta = \eta_o \cdot \exp(Q/RT) \tag{3}$$

where R is the gas constant and η_o is the pre-exponential factor. The activation energy of densification, Q, is the slope of an Arrhenius plot of $\dot{\bar{\rho}}$ vs. 1/T at a given density. In reactive two-phase systems, e.g. the glass phase chemically react with the ceramic particles during the sintering process, the glass viscosity may vary with density giving rise to an apparent activation energy. Jean and Gupta observed[10] this density dependence of Q in a reactive BSG+high silica glass (HSG) system at sintering temperatures above 800°C. However, the BSG+silica system under this study is nonreactive because of the low sintering temperatures used and therefore Q is expected to be independent of density.

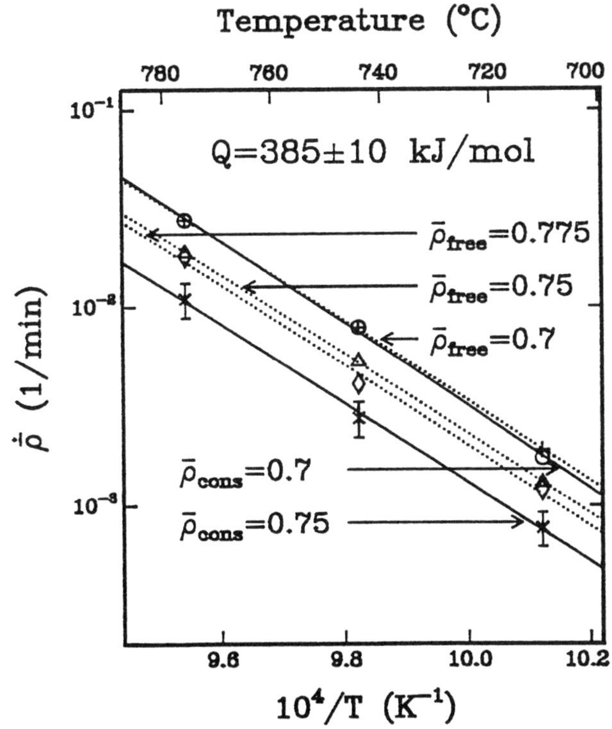

Figure 8. Activation energy plots for free and constrained BSG+Silica films.

Plotted in Figure 8 are $\dot{\bar{\rho}}$ vs. 1/T lines obtained at relative densities between 0.7 and 0.775. These plots are based on $\dot{\bar{\rho}}$ values taken from the free-film and constrained-film densification profiles that were sintered at 715°C, 745°C, and 775°C. The fact that all the lines are nearly parallel suggests a density independent activation energy for densification and an identical mechanism for both free and constrained films; Q was 385 ± 10 kJ/mol.

Without an effect by the substrate constraint on the energetic of densification in the dielectric films, how can one explain the observed rate-retardation in the constrained films? We suggest to consider stresses developed in the constrained-sintering films; the stresses will be tensile and therefore lower the driving force for densification. Recently, we measured[11] the film stresses in pure BSG films on rigid substrates. We found that the film stresses could slow-down or prevent some large-sized pores from shrinking, resulting in less dense films. Work is in progress to measure stresses developed in the BSG+silica films.

SUMMARY

The constrained-film sintering kinetics of a pure gold and BSG+silica films were measured using an optical setup. In both materials, the densification rates in the constrained films were considerably reduced in comparison with those in the free films; SEM micrographs of the sintered films confirmed much more porous microstructures in the constrained films. Much higher sintering temperatures are necessary to produce constrained films as dense as free films. In the gold which densifies via solid state diffusion, the substrate constraint also had an effect on the energetics of densification; it changed the predominate mechanism from grain boundary diffusion in the free films to bulk diffusion in the constrained films. In the BSG+silica, the substrate constraint had no effect on the activation energy for densification which was via a viscous flow mechanism.

ACKNOWLEDGMENT

We thank Mr. Robert C. Sutterlin and Dr. Baoping He for technical assistance and Mr. Weiqun Gu for discussions.

REFERENCES

[1]. P.J. Holmes and R.G. Loasby, *Handbook of Thick Film Technology*, Electrochemical Publications, Ayr, Scotland, 1976.

[2]. R.R.Tummala, in "*Microelectronics Packaging Handbook*", edited by R.R. Tummala and E.J. Rymaszewaki (Van Nostrand Reinhold, New York, 1989), Chap. 7.

[3]. G-Q. Lu, R.C. Sutterlin, and T.K. Gupta, "Effect of Mismatched Sintering Kinetics on Camber in a Low Temperature Cofired Ceramic Package," *J. Am. Ceram. Soc.* **76**, 1907 (1993).

[4]. R.C. Sutterlin, G-Q. Lu, and T.K. Gupta, "Cosintering of Gold/LTCC Substrates: Development of Camber and Via Defects,"*Ceram. Trans.* **33** ,435 (1993).

[5]. J.C. Bang and G-Q. Lu, "Densification Kinetics of Glass Films Constrained on Rigid Substrates," to be published in J.Mater. Res. (1995).

[6]. K.R. Venkatachari and R. Raj., "Shear Deformation and Densification of Ceramic Compacts", *J. Am. Ceram. Soc.*, **69**, [6] 499-506 (1986).

[7]. E.A. Brandes, *Smithells Metals Reference Book*, 6th ed. (Butterworth & Co. Publishers, Ltd., 1983).

[8]. R.K. Bordia and G.W. Scherer, "Sintering of Composites: A Critique of the Available Analyses," *Ceram. Trans.* **1**, 872-886 (1988).

[9]. G.W. Scherer, "Sintering of Low-Density Glasses: I, Theory," *J. Am. Ceram. Soc.* **60**, 236 (1977).

[10]. J-H. Jean and T.K. Gupta, "Densification Kinetics of Binary Borosilicate Glass Composite," *J. Mater. Res.* **9**, 486 (1994).

[11]. J.C. Bang and G-Q. Lu, "Constrained-Film Sintering of a Borosilicate Glass: In-situ Measurement of Film Stresses," to be published in *J. Am. Ceram. Soc. Comm.* (1995).

INFLUENCE OF STRUCTURAL DEFECTS ON THE DIELECTRIC PROPERTIES OF CERAMIC MICROWAVE RESONATORS

Peter K. Davies
Department of Materials Science and Engineering
University of Pennsylvania
3231 Walnut St., Philadelphia, PA 19104-6272

INTRODUCTION

The unique electrical properties of ceramic dielectric resonators have revolutionized the microwave-based wireless communications industry by reducing the size and cost of filter and oscillator components in systems ranging from cellular telephones to global positioning technologies. Microwave resonator ceramics must have a high dielectric constant (typically $\varepsilon_r > 25$), low dielectric loss or high "Q" in the microwave region ($Q = 1/\tan\delta = > 5000$) and a near-zero temperature coefficient of resonant frequency ($\tau_f = 0 \pm 4$ppm/°C). Over the last 20 years several oxides have been found to meet these criteria (Table I) and, for example, ceramics in the BaO-TiO_2 ($Ba_2Ti_9O_{20}$, $BaTi_4O_9$), ZrO_2-SnO_2-TiO_2 ($Zr_{1-x}Sn_xTiO_4$), BaO-Ta_2O_5-ZnO ($(BaZn_{1/3}Ta_{2/3})O_3$ and BaO-PbO-RE_2O_3-TiO_2 (e.g. $Ba_{6-3x}RE_{8+2x}Ti_{18}O_{54}$) systems have found widespread commercial application.[1,2]

TABLE I

System	ε_r	Q (f GHz)	τ_f ppm/°C
$Ba_2Ti_9O_{20}$	40	12,500 (4.5)	3
$BaTi_4O_9$	38	13,000 (4.5)	13
$(Zr_{1-x}Sn_x)TiO_4$	35-40	10,000 (4.5)	0
$Ba_{6-3x}RE_{8+2x}Ti_{18}O_{54})$	80-90	2000 (3)	0
$Ba_3Ta_2MeO_9$ (Me = Zn/Mg/Ni)	28-30	>20,000 (10)	0

While these materials are adequate for current applications, the drive toward further miniaturization and improved filtering capabilities will require the development of new ceramics with higher dielectric constants and lower losses. Although some guidelines do exist, an understanding of the key chemical and structural requirements for a successful ceramic resonator is only in its infancy. This progress has been impeded by the complex crystal chemistry of many of the existing ceramics whose structures are often highly disordered through the formation of a wide variety of extended defects. This first part of this paper will provide a brief review some of the efforts that have been made to correlate the dielectric properties of these oxides to their chemistry and structures. The second part of the paper will focus on some of our own studies where we have used high resolution transmission electron microscopy (HRTEM) to correlate the local structural order in Sn-$ZrTiO_4$ and $BaZrO_3$-$Ba(Zn_{1/3}Ta_{2/3})O_3$ ceramics to their dielectric loss properties. In particular it will be shown that by stabilizing the internal domain boundaries that form in these systems it is possible to significantly enhance the dielectric response.

MICROWAVE CERAMICS: CHEMISTRY-STRUCTURE-PROPERTIES.

The materials listed in Table I have a wide range of stoichiometries, structures, dielectric constants and dielectric losses; given these differences, to what extent is it possible to rationalize or pre-predict the properties of a given resonator material ? In the microwave region the magnitude of the dielectric constant is determined by the ionic and electronic polarization, with the ionic polarization being the most important for the higher dielectric constant oxides. In 1993 Shannon published a set of empirically-derived ionic polarizabilities, α, which enable prediction of the dielectric constant of oxides from the Clausius-Mossotti relationship, where $\alpha = (4\pi/3V)[(\varepsilon - 1)/(\varepsilon + 2)]$ and $\alpha = \sum\alpha_{ions}$, V = molar unit cell volume.[3] Despite the limitations of this empirical approach it works well for many materials with a low dielectric constant but is largely untested for the materials of interest for resonator applications with dielectric constants ranging from 20-150. We have found that this method gives good results for $ZrTiO_4$ and $Ba_2Ti_9O_{20}$ - based dielectrics, but unreliable predictions for Ta-oxides and other perovskite-related systems suggest that further refinements of the polarizabilities are required; it should be noted that other groups are currently working to do this.[4] Nevertheless, Shannon's polarizabilities do serve as a useful guide in the development of oxide dielectrics and in many cases can permit predictions of the dielectric constant knowing only the stoichiometry and unit cell volume.

It is only recently that correlations have been made between the intrinsic structure and temperature coefficients of resonant frequency of oxide dielectrics. Of particular note here is the work of Setter's group in Lausanne who established a direct correlation between the value of the temperature coefficient of relative permittivity and the occurrence of octahedral tilts in $Sr(Zn_{1/3}Nb_{2/3})O_3$-$Ba(Zn_{1/3}Nb_{2/3})O_3$ perovskite solid solutions.[5] Setter has also reported that the values of τf in many perovskite oxides scale with the tolerance factor of the structure.[6] We have found that a similar empirical relationship holds between τf and deviations from ideal hexagonal packing in oxides related to $ZrTiO_4$,[7] and structure-τf correlations are also evident in recent work conducted on the high dielectric constant $Ba_{6-3x}RE_{8+2x}Ti_{18}O_{54}$ systems.[8]

Although some progress has been made in correlating the losses and relative permittivities of a group of $Ba(B_{1/2}B'_{1/2})O_3$ perovskites,[9] relatively few guidelines exist for quantitative pre-prediction of the dielectric loss of oxide dielectrics. Several groups have used Kramers-Kronig analysis and dispersion theory to calculate the intrinsic losses of microwave ceramics from far IR reflectance spectra.[10] In many cases the losses calculated via this approach are in good agreement with those determined via direct measurement, and these studies have provided useful insights into the relative contributions of the individual resonant modes of the lattice to the total loss. In addition to losses from phonon scattering mechanisms, lattice dynamics calculations have provided evidence for a correlation between the charge distribution and microwave losses of crystals, e.g. spinel and perovskite, in which two ions with different oxidation states are located on a single set of lattice sites.[11,12] In both cases as the spatial correlation between the different charged cations is decreased, significant increases in the dielectric loss arise from the coupling of the charge distribution to transverse acoustic modes. For example, Sagala's calculations for $Ba(Zn_{1/3}Ta_{2/3})O_3$ predicted microwave loss tangents (=1/Q) of 10^{-3} and 10^{-6} for disordered and ordered B-site arrangements respectively.[12] Although the calculated Q differences are in good agreement with experimental results, it should be noted that this type of analysis involves the assumption of formal oxidation state charges on the different cations, e.g. 2 and 5 for Zn and Ta; the magnitude of this effect would be significantly reduced if the real charges of the ions, which are typically much lower than those indicated by a formal ionic model, were included in the model.

It is well recognized that losses associated with atomic level defects can dominate the dielectric response of all these systems.[13,14] Because the resonators are ceramics, the formation of defects and their associated losses can arise from several different sources such as: (1) raw material impurities; (2) grain boundary impurities originating from the raw materials or from deliberately added sintering aids which are often essential to achieve high density; and (3) charge carriers and structural defects associated with oxidation/reduction during processing.[2] Negas has also shown that because many of these systems are inherently anisotropic, preferred orientation induced by different processing methodologies can lead to large alterations in the dielectric response.[2] The sensitivity of the dielectric loss to each of these effects has often obfuscated the role of the intrinsic crystal structure. As described below, the crystal structures of almost all the leading oxide resonators exhibit substantial deviations from perfect periodicity and it has become apparent that these internal inhomogeneities can be responsible for large alterations in the dielectric loss.

STRUCTURAL DISORDER IN MICROWAVE CERAMICS.

The studies made by our group using HRTEM[15-20] and by others using x-ray diffraction techniques,[21-24] have demonstrated that the crystal structures of all the prominent microwave ceramics can exhibit extensive internal inhomogeneity. This structural "disorder" arises from the formation of extended defects (e.g. $BaTi_4O_9$, $ZrTiO_4$), polytype intergrowths (e.g. $Ba_2Ti_9O_{20}$) or variable degrees of cationic order (e.g. Sn-$ZrTiO_4$, $Ba(Zn,Mg)_{1/3}Ta_{2/3}O_3$). In several cases it has been demonstrated that alterations in the degree of the intrinsic disorder, through different processing methodologies or chemical substitution, have a pronounced affect on the dielectric properties, in particular the loss.[17,23,24] This structure-loss relationship is perhaps best illustrated by the two orders of magnitude change in Q that results from the long-range ordering of the B-site cations in Ta-based perovskites.[23,24] When the Zn and Ta ions in $Ba(Zn_{1/3}Ta_{2/3})O_3$ are "disordered" the Q factor is <500, while Q values in excess of 35,000 at 10GHz have been measured for samples in which the cations adopt a long-range ordered arrangement on the octahedral sites.

Despite the importance of the crystal chemistry in these systems, it is only recently that the atomic level structures and defects have been characterized and related to the dielectric properties. In our own studies we have found that HRTEM is the technique of choice for understanding the local structures of many of the microwave ceramics. In the late 1980's we published papers on the local structures of the barium polytitanates and demonstrated that these systems are highly "disordered", primarily through the intergrowth of different polytypes and extended defect formation.[19,20] More recently this work was extended to zirconium titanate-based ceramics and currently to the so-called super Q perovskites such as $Ba(Zn_{1/3}Ta_{2/3})O_3$ which have the lowest losses of all the prominent resonators. Our goals have been to apply appropriately sensitive structural analysis methods, principally HRTEM, to understand which features of the crystal chemistry are responsible for alterations in the dielectric properties and to use this information to design and synthesize new resonator materials. The following sections summarize some of the recent work on the $Zr_{1-x}Sn_xTiO_4$ and $BaZrO_3$-substituted $Ba(Zn_{1/3}Ta_{2/3})O_3$ systems.

HRTEM STUDIES OF $(Zr_{1-x})Sn_xTiO_4$.

The oxides of interest for microwave applications in the ZrO_2-TiO_2-SnO_2 system adopt the α-PbO_2 structure above 1200°C with the cations disordered among equivalent octahedral sites.[25,26] Below 1200°C all of the compositions in the ZrO_2-TiO_2 binary system ($Zr_{1-y}Ti_{1+y}O_4$, $0.0 \leq y \leq 0.167$) undergo a continuous phase transformation characterized by increasing order in the Zr-Ti distribution.[22] Ti-rich samples near $ZrTi_2O_6$ (e.g. $Zr_5Ti_7O_{24}$) adopt a structure in which the larger Zr ions order onto every third layer along the *a* axis and distort the adjacent oxygen anions to adopt an approximately 8-fold coordinate site.[26] This commensurate phase has an *a* repeat exactly three times that of the disordered structure with a {..ZrTiTiZr..} repeat sequence. We established that the ordered $ZrTiO_4$ end-member had a {..ZrZrTiTi..} layer sequence and a commensurate 2x repeat along *a*.[15] The intermediate-composition ordered zirconium titanates show incommensurate diffraction effects, however their structures are not modulated in the true sense and instead are comprised of a random mixture of the {..ZrTiTi..} and {..ZrZrTiTi..} modules of the two end-members, see Figure 1.[15,16] While the Zr and Ti cations in the fully ordered structures should occupy sites in individual (100) cation layers; kinetic factors inhibit the complete segregation of the cations and each layer is actually comprised of two-dimensional Zr-rich and Ti-rich domains. This is highlighted by the image in Figure 2a, in which the cation layers undergo changes in occupancy from a Zr-rich to Ti-rich composition, or vice-versa. Typically these “switching boundaries” convert one Zr layer of a {ZrZrTiTi} slab into a Ti layer, and this “new” Ti layer is compensated by the conversion of an adjacent Ti layer to a Zr layer.

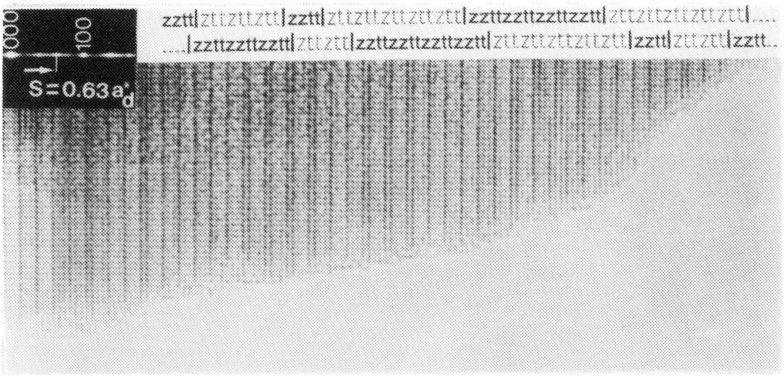

Figure 1; [010] image of $Zr_{0.83}Ti_{1.17}O_4$ - darker rows Zr layers.

To characterize the effect of the replacement of Zr by Sn on cation ordering and the effect of ordering on the dielectric loss, we investigated the crystal chemistry and microwave dielectric properties of $Zr_{1-x}Sn_xTiO_4$ ceramics with x ranging from 0.0 to 0.15.[17] Although the Sn substitutions inhibit ordering, we found that the effect is progressive and ordering still proceeds to an observable degree in samples with x = 0.15. Images collected along the [001] direction again showed that the individual (100) cation layers switch abruptly from Zr-rich to Ti-rich occupancy, Figure 2b, indicating that the ordering proceeds by the segregation of Zr and Ti into elongate two-dimensional domains on (100). However, the size of the ordered domains was found to decrease as the Sn content was increased. In the Sn-free samples a switch in the layer occupancy occurs, on average, every 425Å; for x = 0.09 the boundaries occur every 180Å while the length scale of the cation correlations is approximately 25-50Å for x = 0.15 (Figure 3). These changes appear to be directly related to the stabilization of the Zr/Ti interfaces by the Sn cations, probably through the concentration of Sn at or near the domain boundaries.

The effect of the ordering process on the dielectric properties was examined from microwave measurements on disordered and ordered ceramics of $Zr_{1-x}Sn_xTiO_4$. To ensure that any changes in the properties could be related to the alterations in the intrinsic crystal structure, our processing routes were designed to avoid the deleterious effects associated with the reduction of Ti or grain boundary impurities. To this end we used the sol-gel method for the Sn-free samples and hot-forging techniques for the Sn-doped samples. Whereas the long-range cation ordering increased the dielectric loss of $ZrTiO_4$ by 30%, it resulted in a very small improvement in the loss properties of $Zr_{0.91}Sn_{0.09}TiO_4$, see Table II for a summary complete data appears in [17]. We feel the hypothesized effect of Sn in stabilizing the Zr-Ti domain boundaries may also be responsible for reducing the contribution of these sites to the dielectric loss of the ordered structures.

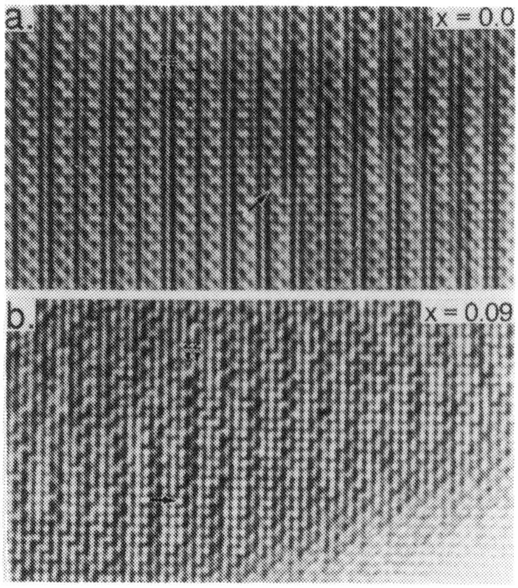

Fig. 2. [001] images of $Zr_{1-x}Sn_xTiO_4$; (a) x=0, (b) x=0.09 domain boundary indicated by arrow

Fig. 3 [001] image of $Zr_{0.85}Sn_{0.15}TiO_4$ inclined view shows ordered domains

TABLE II. Properties of Ordered and disordered $Zr_{1-x}Sn_xTiO_4$

x	Structure	V(Å^3)	ε_r	Q	f(GHz)	Q.f	τ_f (ppm/°C)
0.0	Disordered	132.5	43.1	3122	8.25	25756	59
0.0	Ordered	130.3	39	2171	8.6	18680	77
0.09	Disordered	132.6	38.6	3233	8.7	28127	24
0.09	Ordered	130.7	36	3206	9.3	29710	30

HIGH Q PEROVSKITES: THE $BaZrO_3$ - $Ba(Zn_{1/3}Ta_{2/3})O_3$ SYSTEM

Introduction, background.

For high frequency applications in the 10-30GHz band a series of ceramics with Q factors exceeding 10,000 at 10Ghz were developed in the 1980's.[23,24] Because of the higher frequency application, candidate materials for the resonator components are not required to have particularly high dielectric constants, typically ≤ 35, but a very high Q is mandatory in providing high frequency selectivity ($Q = fo/\Delta f$). Several oxides based on the perovskite structure have been found to satisfy these requirements, though optimization of their loss properties has required careful control of their chemistry, processing and degree of structural order.

The perovskites that exhibit the highest Q values are based on the general formula, $Ba(M^{2+}{}_{1/3}Ta^{5+}{}_{2/3})O_3$, where M = Zn or Mg. Through the work of Galasso and others several years ago, it is well known that many perovskites with a 1:2 mixture of divalent and pentavalent ions on the B-sites adopt a structure in which the two cations order onto individual (111) crystallographic planes.[27-29] The long-range ordering, which is often accompanied by a small lattice distortion, yields a hexagonal structure with a {..M^{2+}-M^{5+}-M^{5+}..} repeat sequence along the <111> direction of the parent cubic perovskite cell, see Figure 4. Studies of the structures and properties of the tantalate systems revealed that the degree of ordering between the M^{2+} and Ta^{5+} ions has a pronounced effect on the dielectric loss at microwave frequencies. By inducing long-range cation order through extended high temperature annealing (≥1400°C), the Q values of $Ba(M^{2+}{}_{1/3}Ta^{5+}{}_{2/3})O_3$ ceramics can be increased from ~500 to >35,000 at 10GHz.[24] Although some investigators initially claimed the improvements in Q were related to Zn-volatilization,[30] several subsequent studies of the Mg and Zn systems in which the volatilization was minimized or eliminated, have clearly demonstrated that the changes in loss arise from alterations in the degree of intrinsic cation order.

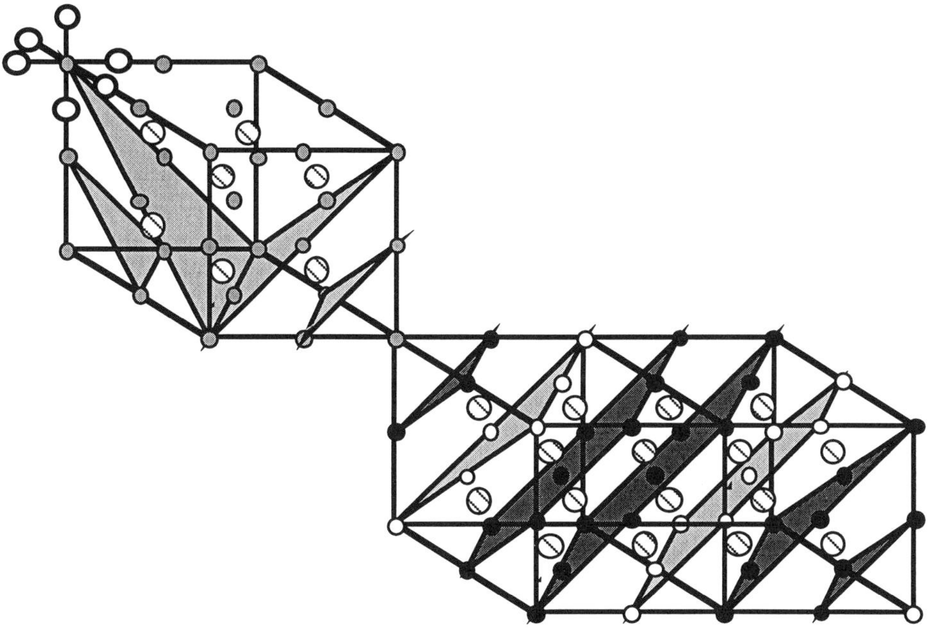

Figure 4. Schematic illustration of cation ordering in $Ba(Zn_{1/3}Ta_{2/3})O_3$; upper left shows two of the possible <111> directions for the ordering of Zn and Ta. Lower part of diagram illustrates the ordering for one set of planes. Black circles Ta, small open circles Zn, hatched circles Ba. Oxygens are only included at the top left to illustrate the octahedral coordination of the Zn/Ta ions.

While the fully ordered barium-tantalum-based perovskites have outstanding losses and can be tuned to a zero τf, for example by partial substitution of the Zn ions in

$Ba(Zn_{1/3}Ta_{2/3})O_3$ (τf=3ppm/°C) by Ni (for $Ba(Ni_{1/3}Ta_{2/3})O_3$ τf = -17ppm/°C), from a commercial perspective they are not suited for large-scale production because of the long annealing times required to reach a fully ordered cation distribution. At 1400°C the high Q ordered perovskite ceramics require at least 130 hours annealing. Wakino showed this problem could be circumvented through minor additions of $BaZrO_3$, typically ≤ 5 mole %, which reduce the annealing times, ≤ 10 hours, required to reach the high Q state.[31] Therefore, state-of-the-art high Q perovskites are actually formulated from a mixture containing approximately 89% $Ba(Zn_{1/3}Ta_{2/3})O_3$, 7.5% $Ba(Ni_{1/3}Ta_{2/3})O_3$ and 3.5% $BaZrO_3$ and have ε_r = 29, Q >12,000 at 10 GHz and τf=0.[2]

Although the studies of pure $Ba(M_{1/3}Ta_{2/3})O_3$ (M=Zn,Mg,Ni) perovskites seem to yield a consistent picture of the relationship between the cation ordering, the associated structural distortions, and the dielectric loss, the behavior of the $BaZrO_3$-doped systems is not as straightforward. In his report on the large reduction in the annealing times required to reach a high Q state, Wakino examined the microwave characteristics and cell geometries of $Ba(Zn_{1/3}Ta_{2/3})O_3$ - $BaZrO_3$ and $Ba(Zn_{1/3}Ta_{2/3})O_3$ - $SrTiO_3$ solid solutions.[31] For the substitution of $BaZrO_3$ the high Q values were maintained for additions up to approximately 4 mole % and then degraded gradually at higher concentrations. In the $SrTiO_3$ system the Q's deteriorated rapidly for all concentrations. It was also noted that both substitutions accelerated the "crystallization" of the perovskite structure and the low $BaZrO_3$-doped samples yielded high Q ceramics after very short annealing times (4 hours at 1500°C). Perhaps the most surprising result, particularly in light of the well documented improvement of Q with increased cation order in the un-doped tantalate perovskites, was the observation that the x-ray lines associated with the long-range ordering had disappeared in the 4% $BaZrO_3$ high Q samples. The low concentration additions of $SrTiO_3$ had a similar effect on the structural order, though in this case the Q's were inferior to the un-doped samples. The difference in the effect of the two substitutions on the dielectric properties was attributed to the lattice constants of the added components. It was suggested that because the lattice constant of $BaZrO_3$ is larger than $Ba(Zn_{1/3}Ta_{2/3})O_3$, the improvement in Q arises from the stronger bonding of the Zr ions in the solid solution, whereas the introduction of a substituent with a smaller lattice constant, such as $SrTiO_3$, degrades the Q because the Ti ions are more loosely bound.[31]

Ordering in the $Ba(Zn_{1/3}Ta_{2/3})O_3$-$BaZrO_3$ System.

The high Q values reported by Wakino for the $Ba(Zn_{1/3}Ta_{2/3})O_3$-$BaZrO_3$ system have been reproduced by several groups and these additions are currently used to produce commercial resonators. However, the origins of the improved Q's, the explanations for the decrease in the annealing time required to reach a high Q state, and the observation of an apparently disordered structure with very small levels of $BaZrO_3$, have not been explained. Furthermore, the formation of a high Q ceramic with a random cation arrangement contradicts the experimental and theoretical studies of the un-doped tantalate systems. Despite their importance, no detailed structural studies of the Zr-substituted perovskites have been made.

Before discussing the results of our HRTEM studies on this pseudo-binary system, it is useful to re-consider the ordering process in the pure $Ba(M_{1/3}Ta_{2/3})O_3$ systems. When these phases are first prepared, they crystallize in an apparently "disordered" cubic structure. With extended annealing the "degree of order" (as gauged by the increase in the intensity and sharpness of the superlattice lines in the x-ray patterns) increases and as a "completely ordered" structure is approached the unit cell undergoes an hexagonal distortion. Some authors have implied this transformation is second order and involves a gradual and continuous segregation of the two cations to their respective sites. However, it is more likely that a correct description of this irreversible ordering process, as pointed out in the first studies made by Galasso,[27,28] is in terms of the nucleation and growth of small ordered domains with increasing annealing time and temperature. The formation of this type of domain structure necessarily implies that "disordered" or "partially ordered" samples contain high levels of unstable domain boundaries. Two types of domain boundary are likely to be produced during the ordering of the B-site cations. The first is a twin boundary separating domains in which the nucleation and growth of the ordered structure has propagated along different <111> orientations - three discrete variants are possible, two are shown in Figure 4. The second is an anti-phase boundary associated with each set of ordering orientations. From an energetic and structural viewpoint, these boundaries are highly unfavorable as they produce a region with local cation disorder and considerable associated structural strain. We suggest that the low Q values reported for disordered or partially ordered samples of $Ba(M_{1/3}Ta_{2/3})O_3$ result from high losses associated with the domain boundaries and that the improved losses of "fully ordered" ceramics originate from the growth and eventual formation of single-domain grains.

In parallel with our observations on Sn-substituted $ZrTiO_4$ ceramics, we felt it was possible that the improved losses and shortened annealing times in the $BaZrO_3$-$Ba(Zn_{1/3}Ta_{2/3})O_3$ system could result from the stabilization of domain boundaries by the preferential segregation of Zr. If Zr^{4+} is indeed concentrating at or near the boundaries, the size of the domains would be expected to decrease as the concentration of $BaZrO_3$ is increased. Our preliminary HRTEM results for this system are presented below.

HRTEM Studies of $(1-x)\{Ba[(Zn_{1-y}Ni_y)_{0.33}Ta_{0.67}]O_3\}$ - $x\{BaZrO_3\}$ Resonators.

A series of commercially produced $(1-x)\{Ba[(Zn_{1-y}Ni_y)_{0.33}Ta_{0.67}]O_3\}$ - $x\{BaZrO_3\}$ ceramics with well-defined chemistries and microwave properties (see Table III) were provided by Dr. T. Negas (Transtech, Inc) for examination by HRTEM. Each ceramic was processed to a high Q state; however the annealing times required to access the high Q state were reduced as the concentration of Zr increased (140 hours for the lowest concentration, 8 hours for the highest). As the Zr content of these samples is increased, the ratio of Zn:Ni is systematically raised to maintain τf close to zero. Because the crystal chemistry of Ni and Zn is so similar in this system, we believe the small alterations in the relative concentrations of the two divalent ions does not have any significant effect on the results described below.

TABLE III

Sample #	Composition	Q (10GHz)	εr	τf(ppm/°C)
D8731	$Ba_3(Zr_{.0375}Ni_{.1975}Zn_{.79}Ta_{1.975})O_9$	11448	28.2	-5
D8732	$Ba_3(Zr_{.0645}Ni_{.1625}Zn_{.816}Ta_{1.957})O_9$	12686	28.7	-2
D8733	$Ba_3(Zr_{.09}Ni_{.125}Zn_{.845}Ta_{1.94})O_9$	10018	29.3	1
D8734	$Ba_3(Zr_{.1275}Ni_{.0725}Zn_{.885}Ta_{1.915})O_9$	11909	29.7	2
D8735	$Ba_3(Zr_{.15}Ni_{.0375}Zn_{.9125}Ta_{1.900})O_9$	10634	30.7	4

The HRTEM investigations of the samples listed in Table III were made using a Philips 430 and JEOL 4000EX electron microscope; x-ray diffraction data was collected using a Rigaku DMAXB diffractometer. The x-ray powder pattern of sample D8731 (1.25% $BaZrO_3$) contained additional reflections from a hexagonal superstructure and also had a c/a ratio indicative of a distorted unit cell. The superstructure reflections in sample D8732 (2.15% $BaZrO_3$), were extremely weak and the cell showed no deviation from cubic symmetry. The x-ray patterns of the samples containing 3, 4.25 and 5% $BaZrO_3$ could all be indexed on an ideal cubic cell and showed no evidence for any cation order.

Electron diffraction patterns collected from the first sample (1.25% $BaZrO_3$) along the [110] direction of the pseudo-cubic sub-cell, contained strong additional reflections corresponding to a tripled cell along one unique <111> direction. Lattice images of this sample indicated that the structure was close to perfectly ordered, occasional twin boundaries were observed in some grains though these were few and far between. The images and diffraction patterns collected from samples containing 2.15% $BaZrO_3$ revealed a quite different situation. Superlattice reflections in the [110] diffraction patterns were observed along both of the allowed <111> directions (Figure 5) implying that the grain is comprised of a twinned domain structure. High resolution images, Figure 6, collected from the same area of the sample confirm this interpretation and provide direct evidence for the formation of ordered domains with a twin relationship. It is clear that this sample contains domains within which the cations are ordered along each of the three possible <111> orientations of the cubic sub-lattice. The two <111> directions allowed in this orientation are indicated in Figure 6. The average size of the domains in this image is approximately 400Å.

The diffraction patterns collected from the 3%$BaZrO_3$ sample (which is disordered to x-rays) also show evidence for cation order. The superlattice reflections are significantly weaker and more diffuse, but again are consistent with the formation of a twinned domain structure. Superlattice fringes oriented along the two <111> directions of the cubic sub-cell can be observed in the image in Figure 7, which shows the ordered domains are significantly smaller, approximately 50Å, than in the previous sample. At this stage we have no direct evidence that the reduction in the size of the domains is associated with the "wetting" of the boundaries by the Zr ions. However, the direct correlation between the domain size and the Zr content would be consistent with that interpretation.

Fig. 5. [110] zone axis diffraction pattern of D8732 (2.15%$BaZrO_3$); note superstructure reflections occur along both of the in-plane <111> directions.

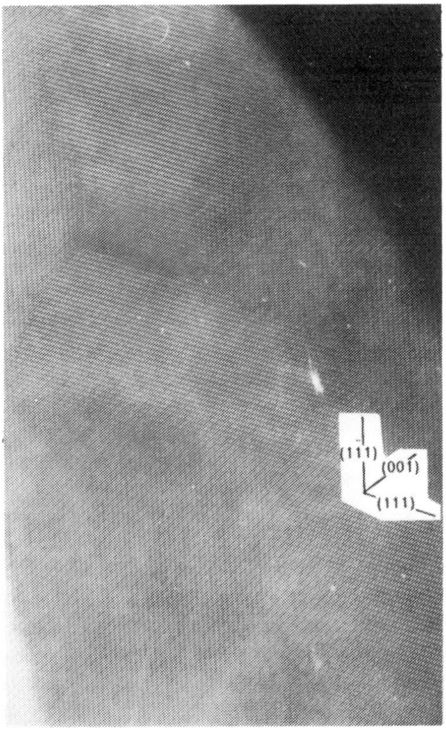

Fig. 6. Image of D8732 (2.15% $BaZrO_3$) along [110]. Note superlattice fringes from twin domains which are ≈ 400Å.

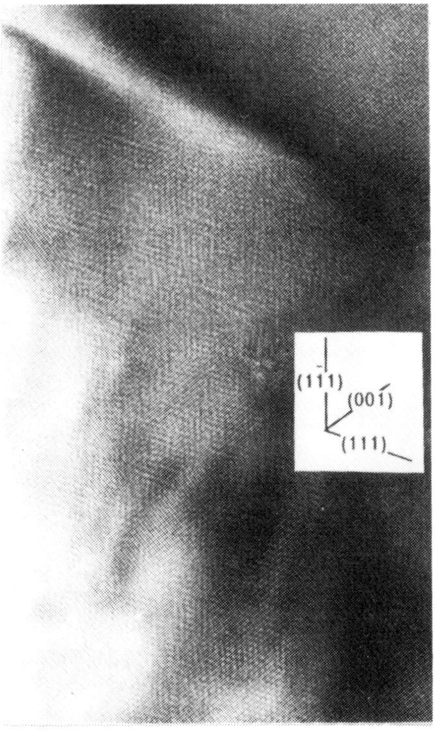

Fig. 7. Image of D8733 (3%$BaZrO_3$) Additional fringes from ordering Domain size ≈ 25-50Å.

(1-x){$Ba[(Zn_{1-y}Ni_y)_{0.33}Ta_{0.67}]O_3$} - x{$BaZrO_3$}: Ordered Domain Structures

From the HRTEM results presented above it is clear that the $BaZrO_3$-doped $Ba(Zn_{1/3}Ta_{2/3})O_3$ high Q perovskites contain an ordered arrangement of Zn(Ni) and Ta on the octahedral lattice sites. Furthermore, the large reduction in the length-scale (domain size) of the cation correlations with increasing concentration of $BaZrO_3$ is apparently responsible for the reduction in the annealing times required to access a high Q state. Without the addition of $BaZrO_3$, the losses of a pure $Ba(Zn_{1/3}Ta_{2/3})O_3$ ceramic with a "partially ordered" structure containing high concentrations of domain boundaries are very high. We propose that the low losses induced by the addition of $BaZrO_3$ result from the stabilization of the domain boundaries. We also suggest that the reduction of the losses at these regions is directly related to the reduction in the degree of strain at the interface and arises from the preferential segregation of the Zr ions to the boundary regions. However, it must be noted that our evidence for the segregation is only that inferred from the response of the structure and properties to the introduction of the Zr^{4+} ions. Direct chemical analyses of these interfaces has not, as yet, been obtained but is being pursued using STEM techniques.

The role we propose for the tetravalent dopants in these ordered perovskites is one in which the M^{4+} cations preferentially occupy sites at the twin/domain boundaries and eliminate unfavorable coulombic interactions and associated structural strains that would otherwise result from a locally disordered arrangement of the 2^+ and 5^+ ions. For the substitution to be effective it is presumably mandatory that the added component has a perovskite structure, and likely that the oxidation state of the B-site dopant ions must equal the mean valence of the octahedral lattice sites in the un-substituted perovskite. The size of the tetravalent ions is also likely to be important. The ionic radius of Zr^{4+} (r_{oct} = 0.72Å) lies between those of Zn^{2+}(r_{oct} = 0.74Å) and Ta^{5+} (r_{oct} = 0.64Å) which could be beneficial in the forming a stable, interfacial "buffer layer" between the twinned domains. For these reasons Sn^{4+} should also be an effective substituent (r_{oct} = 0.69Å); however, the smaller size of Ti^{4+} (r_{oct} = 0.605Å) may inhibit its use as an effective boundary segregant. Indeed it should be recalled that $SrTiO_3$ was not effective in improving the loss properties of the barium zinc-tanalates, though this may not necessarily be due to the chemistry of Ti^{4+} but could also be a result of additional structural strains arising from the partial replacement of Ba by Sr on the A sites of the lattice.

CONCLUSIONS.

One of the barriers to the design of new resonator materials has been the lack of detailed information on the relationship between the chemistry, structures and properties of the existing commercial ceramic resonators. This progress has been impeded by the complex crystal chemistry of these systems which typically exhibit substantial deviations from perfect structural homogeneity. However, progress is being made through the development of empirical ionic polarizabilities, and identification of the relationship between τf and Q and the crystal chemical features of individual compounds. In our work we have found that the formation of internal boundaries in systems involving a cation order-disorder reaction (e.g. $Sn-ZrTiO_4$

and $BaZrO_3$-$Ba(Zn_{1/3}Ta_{2/3})O_3$) can have a pronounced effect on the dielectric loss properties. In the absence of secondary dopants, the lattice strains associated with these interfaces can raise the losses by as much as two orders of magnitude. However, the contributions of the interfaces can apparently be controlled, and perhaps eliminated, by introducing selected dopants that segregate and stabilize the local boundary regions. Using appropriate chemical substitution schemes it may be possible to reduce the boundary strains and improve the dielectric losses of several other candidate resonator systems.

ACKNOWLEDGEMENTS.

The results cited in this paper have been collected in collaboration with members of my research group, including Dr. R. Christoffersen, Dr. J. Tong and X. Wei. I would also like to acknowledge Dr. T. Negas at Trans-Tech Inc and Dr. R.S. Roth at NIST who have been constant collaborators throughout many aspects of this research. This work was supported by the National Science Foundation under Grant No. DMR 92-00800 and through the MRL program under grant No. DMR 91-20688.

REFERENCES

1. K. Wakino, K. Minai, and H. Tamura, "Microwave Characteristics of $(Zr,Sn)TiO_4$ and BaO-PbO-Nd_2O_3-TiO_2 Dielectric Resonators", J. Am. Ceram. Soc., 67[4], 278-81 (1984)
2. T.Negas, G. Yeager, S. Bell and R. Amren, "Chemistry and Properties of Temperature Compensated Microwave Dielectrics", in "Chemistry of Electronic Ceramic Materials, Eds. P.K. Davies and R.S. Roth, NIST, SP 804, 21-38 (1990)
3. R.D. Shannon, "Dielectric Polarizabilities of Ions in Oxides and Fluorides", J. Appl. Phys., 73(1), 348-366 (1993)
4. V.J.. Fratello and C.D. Brandle, “Calculation of Dielectric Polarizabilities of Perovskite Substrate Materials for High Temperature Superconductors”, J. Mater. Res., 9(10), 2554-60 (1994)
5. E.L. Colla, I.M. Reaney and N. Setter, "Effect of Structural Changes in Complex Perovskites on the Temperature Coefficient of the Relative Permittivity", J. Appl. Phys., 74(5), 3414-25 (1993)
6. I.M. Reaney, E.L. Colla and N. Setter, "Dielectric and Structural Characteristics of Ba- and Sr-based Complex Perovskites as a Function of tolerance Factor", Jpn. J. Appl. Phys., 33(7A), 3984-90 (1994)
7. X. Wei, P.K. Davies and T. Negas, “Phase Equilibria and Dielectric Properties of Aliovalent Doped Zirconium Titanate Ceramics”, J. Amer. Ceram. Soc., (in preparation)
8. T. Negas and P.K. Davies, these proceedings.
9. A. Tangantsev, J. Petzelt and N. Setter, "Relation Between Intrinsic Microwave and Submillimeter Losses and Permittivity in Dielectrics", Solid State Commun., 87, 1117-20 (1993)
10. See for example: K. Wakino, M. Murata and H. Tamura, "Far Infra-Red Spectra of $Ba(Zn,Ta)O_3$-$BaZrO_3$ Dielectric Resonator Material", J. Am.

Ceram. Soc., 69(1) 34-37 (1986); K. Fukuda, R. Kitoh and I. Awai, "Far infra-red Reflection Spectra of Dielectric Ceramics for Microwave Applications", J. Am. Ceram. Soc., 77(1) 149-541 (1994)

11. E. Schloemann, "Dielectric Losses in Ionic Crystals with Disordered Charge Distributions", Phys. Rev., 135 [2A] A413-19 (1960)
12. D. A. Sagala and S. Nambu, "Microscopic Calculation of Dielectric Loss at Microwave Frequencies for Complex Perovskite $Ba(Zn_{1/3}Ta_{2/3})O_3$", J. Am. Ceram. Soc., 75(9) 2573-75 (1992)
13. A.K. Tagantsev, "Phonon Mechanisms of Intrinsic Dielectric Loss in Crystals", Ferroelectric Ceramics, 127-145 (1993)
14. V.L. Gurevich and A.K. Tangantsev, "Intrinsic Dielectric Loss in Crystals", Adv. Phys., 40, 719-767 (1991)
15. R. Christoffersen and P.K. Davies, "Structure of Commensurate and Incommensurate Ordered Phases in the System $ZrTiO_4$ - $Zr_5Ti_7O_{24}$", J. Am. Ceram. Soc., 75 [3], 563-69 (1992)
16. R. Christoffersen and P.K. Davies, "Extended Defect Intergrowths in $Zr_{1-x}Ti_{1+x}O_4$", Solid State Ionics, 57, 59 (1992)
17. R. Christoffersen, P.K. Davies, X. Wei and T. Negas, "Microstructure and Properties of Sn-doped $ZrTiO_4$ Microwave Ceramics", J. Amer. Ceram. Soc., 77(6), 1441-51 (1994)
18. P.K. Davies and R. Christoffersen, "Influence of Structural Defects on the Properties of Zirconium Titanate-Based Microwave Ceramics", Proc. British Ceram. Soc., 52, 13-27 (1994)
19. P.K. Davies and R.S. Roth, "Defect Intergrowths in Barium Polytitanates 1. $Ba_2Ti_9O_{20}$", J. Solid State Chem., 71, 490-502 (1987)
20. P.K. Davies and R.S. Roth, "Defect Intergrowths in Barium Polytitanates 1. $BaTi_5O_{11}$", J. Solid State Chem., 71, 503-12 (1987)
21. A.E. McHale and R.S. Roth, "Investigation of the Phase Transition in $ZrTiO_4$ and $ZrTiO_4$-SnO_2 Solid Solutions", J. Am. Ceram. Soc., 66 [2], C-18-20 (1983)
22. A.E. McHale and R.S. Roth, "Low-temperature Phase Relationships in the System ZrO_2-TiO_2", J. Am. Ceram. Soc., 69 [11], 827-32 (1986)
23. S. Kawashima, M. Nishida, I Ueda and H Ouchi, "$Ba(Zn_{1/3}Ta_{2/3})O_3$ Ceramics with Low Dielectric Loss at Microwave Frequencies", J. Am. Ceram. Soc., 66 [6], 421-23 (1983)
24. K. Matsumoto, T. Hiuga, K. Takada, and H. Ichimura, "$Ba(Mg_{1/3}Ta_{2/3})O_3$ Ceramics with Ultra-Low Loss at Microwave Frequencies", pp. 118-21 in Proceedings of the 6th IEEE International Symposium on Application of Ferroelectrics, June 1986. Institute of Electrical and Electronic Engineers, New York, 1986.
25. R.E. Newnham, "Crystal Structure of $ZrTiO_4$", J. Am. Ceram. Soc., 50 [4], 216 (1967)
26. P. Bordet, A.E. McHale, A. Santoro, and R.S. Roth, " Powder Neutron Diffraction Study of $ZrTiO_4$, $Zr_5Ti_7O_{24}$, $FeNb_2O_6$", J. Solid State Chem., 64, 30-46 (1986)
27. F. Galasso and J. Pyle, "Ordering in Compounds of the $A(B'_{0.33}Ta_{0.67})O_3$ Type", Inorg. Chem., 2(3), 482-84 (1963)

28. F. Galasso and J. Pyle, "Preparation and Study of Ordering in $A(B'_{0.33}Nb_{0.67})O_3$ Perovskite-Type Compounds", J. Phys. Chem., 67, 1561-62 (1963)
29. A.J. Jacobson, B.M. Collins and B.E. F. Fender, "A Powder Neutron and X-ray Diffraction Determination of the Structure of $Ba_3Ta_2ZnO_9$: An Investigation of Perovskite Phases in the System Ba-Ta-Zn-O and the Preparation of $Ba_2TaCdO_{5.5}$ and $Ba_2CeInO_{5.5}$", Acta Cryst., B32, 1083-87 (1976)
30. S.B. Desu and H.M. O'Bryan, "Microwave Loss Quality of $BaZn_{0.33}Ta_{0.67}O_3$ Ceramics", J. Am. Ceram. Soc., 68(10) 546-51 (1985)
31. H. Tamura, T. Konoike, Y. Sakabe and K. Wakino, "Improved High Q Dielectric Resonator with Complex Perovskite Structure", J. Am. Ceram. Soc., 67(4) C 59-61 (1984)

EFFECTS OF DOPANTS AND OXYGEN VACANCIES ON THE DIELECTRIC LOSS TANGENT OF $(Zr_{0.8}Sn_{0.2})TiO_4$ DIELECTRIC CERAMICS

Yukio Higuchi, Nobuhiko Michiura, Tsutomu Tatekawa and Hiroshi Tamura
Murata Manufacturing Co., Ltd.
Yokaichi-shi, Shiga, 527,Japan

ABSTRACT

The effects of dopants and oxygen vacancies on dielectric loss tangent of $(Zr_{0.8}Sn_{0.2})TiO_4$ materials were investigated. In the case that the specimen was doped with 0.5wt% Ta_2O_5 and then annealed in reduced atmosphere, the increase of loss tangent was enormous at frequencies lower than 100 KHz but slight at microwave frequencies. When the specimen was doped with 0.5 wt% Fe_2O_3, the loss tangent increased drastically at microwave frequencies. These experimental results are well explained by the contribution of space charge polarization and lattice defects with charge deviations.

INTRODUCTION

The $(Zr_{0.8}Sn_{0.2})TiO_4$ material is known as a material suitable for dielectric resonators with superior microwave characteristics of dielectric constant K=38, $Q=1/\tan\delta=8000$ at 7GHz and temperature coefficient of resonant frequency $\tau_f = 0$ ppm/℃.[1]

The crystal structure of this material has been reported by several authors. Newnham determined the crystal structure of $ZrTiO_4$ to be the type of $\alpha-PbO_2$, space group Pbcn.[2] Wolfram and Göbel reported the detailed phase diagram of the system $ZrO_2-TiO_2-SnO_2$.[3] McHale and Roth reported the phase transition of $ZrTiO_4$ around

1130℃.[4] This phase transition was clarified to be the transformation from α-PbO_2 type high temperature structure with a random distribution of Zr and Ti ions to the low temperature structure with commensurate or incommensurate ordered phase of cations.[5,6]

At microwave frequencies, the intrinsic dielectric loss tangent of ionic crystals is determined by the anharmonic terms of the interatomic potential energy of a crystal and is increased by various lattice defects. Schlöman presented a theory of dielectric losses in non-conducting ionic crystals containing disordered charge distributions. Sagala et al. reformulated the theory and applied it to the dielectric loss for the complex perovskite $Ba(Zn_{1/3}Ta_{2/3})O_3$. Their results show the dielectric loss tangent at microwave frequencies depends strongly on the degree of order of cations, which is described in terms of the spatial correlation between the charge deviations at given lattice sites.

In pure $(Zr_{0.8}Sn_{0.2})TiO_4$ crystals, the degree of disorder of cations (Zr^{4+}, Sn^{4+} and Ti^{4+}) does not cause a charge deviation. From the above background, the effects of dopants and oxygen vacancies on the dielectric loss tangent of $(Zr,Sn)TiO_4$ were investigated at frequencies from 0.1KHz to 9GHz. The experimental results distinguish the contribution of the lattice defects with a charge deviation from that of the space charge polarization on dielectric loss tangent.

EXPERIMENTAL PROCEDURE

Four kinds of ceramic specimens A to D with the basic composition of $(Zr_{0.8}Sn_{0.2})TiO_4$ were prepared using the reagent-grade raw material. 0.5wt% ZnO and 0.2wt% NiO were added as sintering aids to all specimens. These divalent ions do not concentrate in grains but form the boundary phases such as $(Zn,Ni)_2TiO_4$.

Furthermore, specimens B and C were doped with 0.5wt% Ta_2O_5 (i.e., 0.47mol% $TaO_{5/2}$), and specimen D was doped with 0.5wt% Fe_2O_3 (i.e., 1.31mol% $FeO_{3/2}$). As the base crystal $(Zr,Sn)TiO_4$ is made of the tetravalent cations, pentavalent Ta ions work as donors and trivalent Fe ions as acceptors.

The raw materials were milled in distilled water with organic binder, dried and pressed into the specimens of cylindrical shape with the dimension of 60mm outer diameter, 30mm inner diameter and 45mm height. All specimens were sintered at

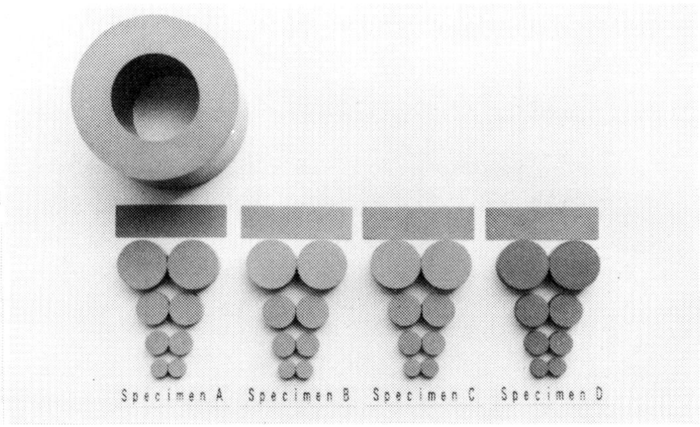

Fig.1. Various size of specimens for measurement use.

1380℃ for 4hrs and cooled slowly. Only specimen C was subsequently annealed at 1100℃ for 2hrs in reduced atmosphere with N_2 of 100L/hr and H_2 of 120L/hr.

The dielectric properties of these specimens were measured at 800MHz. Then, the original specimens were cut into the rectangular specimens with the dimensions of 45×13×0.5mm and the smaller cylindrical samples with the various sizes. Finally they were finished by grinding their surfaces. (See Fig.1)

Using the rectangular specimens with In–Ga electrodes, the insulation resistance was measured by TR8601 High Megohm Meter from Takeda Riken and dielectric properties at frequency range from 0.1KHz to 1MH were measured by HP4284A LCR Meter from Hewlett Pakard. Dielectric properties at microwave frequencies were measured by the dielectric rod resonator method[9] using the cylindrical samples. The measurement instruments were HP8757A Network Analyzer and HP8341A Synthesized Sweeper.

Lattice parameters of test specimens were determined by the X–ray powder diffraction analysis. The Ta and Fe ions' concentration in grains were investigated by transmission electron microscope JEM–2000–FX from JEOL and energy dispersive X–ray spectrometer TN–5000 from Noran Instrument.

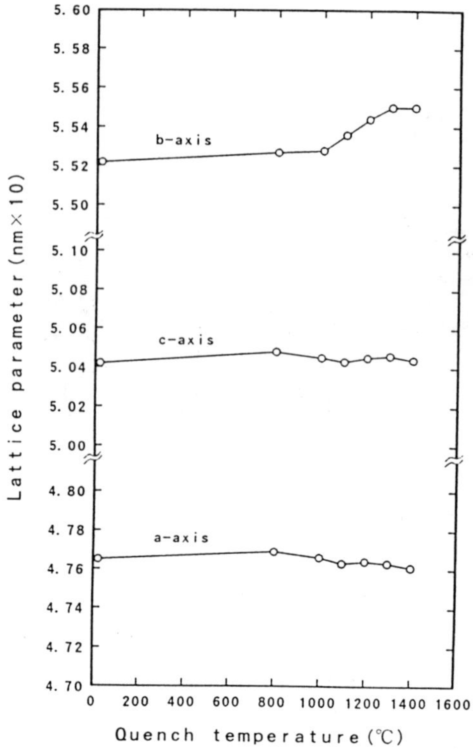

Fig.2. Continuous variation of the lattice parameters of quenched $(Zr,Sn)TiO_4$ specimens

RESULTS

Crystal Structure of Test Specimens

$ZrTiO_4$ was reported to have a phase transition around 1130℃ from the high-temperature phase with a disordered distribution of Zr and Ti ions to a low temperature phase with commensurate or incommensurate ordered arrangement of cations. The phase transition of $(Zr,Sn)TiO_4$ was investigated. The specimen A was cut into small pieces and each piece was fired at various temperature from 800 to 1400℃, which was subsequently

Table I. Lattice Parameters and Electrical Resistivities of $(Zr_{0.8}Sn_{0.2})TiO_4$ Specimens at Room Temperature

	lattice parameters (nm)* a	b	c	resistivities ρ[Ω·cm]
Specimen A	0.477	0.504	0.552	5.9×10^{13}
Specimen B	0.477	0.504	0.552	4.8×10^{13}
Specimen C	0.477	0.504	0.553	5.6×10^{8}
Specimen D	0.477	0.504	0.553	2.0×10^{10}

*specimens cooled at 1°C/min from 1380°C to R.T.

dropped into the water bath. As Fig.2 shows, the continuous variation of the lattice parameters were observed as a function of quench temperature. The lattice parameter of the b axis changes from 0.552 to 0.555nm around 1200℃. Although this difference is much smaller than that in the case of $ZrTiO_4$, whose change was from 0.535 to 0.550nm,[4] the result seems to certify the existence of a minor phase transition on $(Zr,Sn)TiO_4$.

Table I shows the lattice parameters of specimens A–D. All specimens have the same values which coincide with those of the low–temperature phase. As these specimens were cooled slowly with the rate of 1℃/min, it seems reasonable that they have the low temperature phase.

Insulation Resistance

Table I also shows the insulation resistance of the four kinds of specimens. Specimens A and B, which have the same order resistance of 10^{13} Ω•cm, show that the resistance is not decreased by the addition of Ta ions as donor. The resistance of specimen C decreases drastically when the Ta doped specimen is annealed in reduced atmosphere. While, the resistance is decreased by the addition of Fe ions as acceptor.

Dielectric properties at low frequencies

Fig.3 shows the K and tan δ at low frequencies. Both the K and tan δ for specimens C and D increase at lower frequencies, which could be explained by the space charge polarization caused by the decreased insulation resistance.

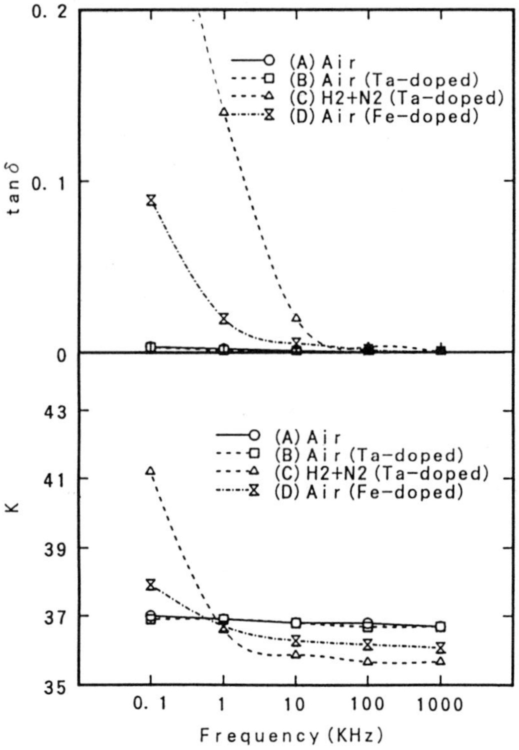

Fig.3. K and tan δ at the frequencies from 0.1 KHz to 1 MHz for the specimens A: no dopant, B: 0.47mol% $TaO_{5/2}$, C: 0.47mol% $TaO_{5/2}$, reduced, and D: 1.31mol% $FeO_{3/2}$

Dielectric properties at microwave frequencies

Fig.4 shows the K and tan δ at microwave frequencies. Significant differences are not observed in dielectric constants but are observed in dielectric loss tangents.

Specimens A and B show the equivalently low tan δ which increases proportionally to frequency. It is worth noticing that the increase of tan δ for specimen C is not serious but slight at microwave frequencies in spite of its low insulating resistance of $10^8 \Omega \cdot$cm. Meanwhile, the Fe doped specimen D shows the drastic increase of tan δ. The interpretations for these experimental results are discussed in the next section.

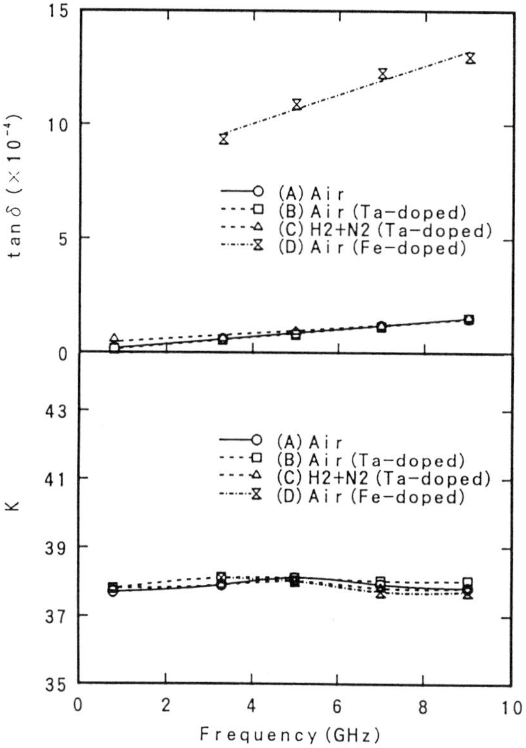

Fig.4. K and tan δ at microwave frequencies for the specimens A: no dopant, B: 0.47mol%$TaO_{5/2}$, C: 0.47mol% $TaO_{5/2}$, reduced, and D: 1.31mol% $FeO_{3/2}$

DISCUSSION

Origins of dielectric loss tangent

As shown in Fig.5, there are three kinds of polarization in paraelectric ionic crystals; space charge polarization, ionic polarization and electronic polarization. At microwave frequencies, the ionic and electronic polarization are dominant and the dielectric properties of ionic polarization are known to be determined by the following dielectric dispersion equation.

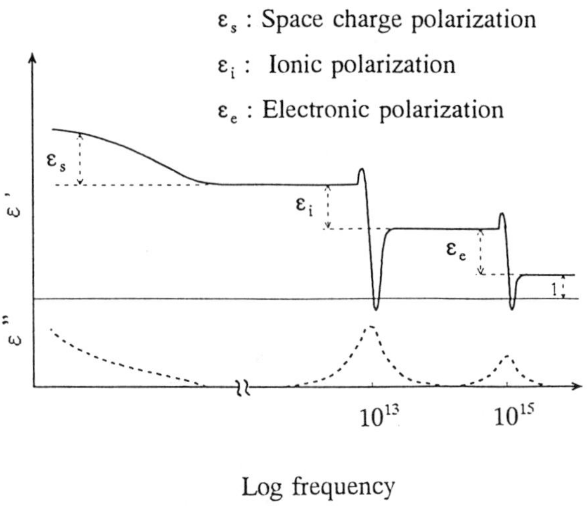

Fig.5. Frequency dependence of the polarization mechanisms in dielectric

$$\dot{\varepsilon}(\omega) - \varepsilon_\infty = \frac{\omega_T^2 (\varepsilon(0) - \varepsilon_\infty)}{\omega_T^2 - \omega^2 - i\gamma\omega} \tag{1}$$

Where ω_T is the frequency of lattice vibration of transverse optical mode, γ is a damping constant of equation of motion of ions, $\dot{\varepsilon}(\omega) - \varepsilon_\infty$ is a relative dielectric constant caused by ionic polarization and $\varepsilon_\infty - 1$ is a relative dielectric constant caused by electronic polarization. As ω_T is around 10^{12} or 10^{13}Hz of infra red region, microwave frequency ω is much less than ω_T. Using the relation $\omega \ll \omega_T$, we obtain the following approximate equations from real part and imaginary part of equation (1).

$$\varepsilon'(\omega) - \varepsilon_\infty \simeq \varepsilon'(0) - \varepsilon_\infty \; ; \; constant \tag{2}$$

$$\tan\delta = \frac{\varepsilon''}{\varepsilon'} \simeq \frac{\gamma\omega}{\omega_T^2} \tag{3}$$

From the above equations, it is found that the dielectric constant doesn't depend on frequency and the dielectric loss tangent is proportional to frequency in the microwave region. ω_T is the constant on the specific material but γ is affected by various kinds of factors. The intrinsic origin of γ is the anharmonic terms of the interatomic potentials which exist even in perfect crystals. The other origins arise from various kinds of lattice defects, which can include
i) grain boundaries and voids ii) order–disorder of ions iii) impurities as donor and acceptor ions and associated vacancies. The first kind of defect such as grain boundaries, boundary phases and voids do not give a serious effect on dielectric loss tangent unless their volume ratio is large. So far, this has been confirmed by the several authors.[1,10]

The second kind of defect, i.e. the degree of order of ions with different valence charge, was found to have a drastic influence on the dielectric loss tangent at microwave frequencies.

Schölmann and Sagala et al. theoretically estimated the dielectric loss tangent for non–conducting ionic crystals with disordered charge distribution on the basis of lattice dynamics. According to the formulation of Schlöman's theory, the dielectric loss tangent is expressed by the following equation.

$$\langle \tan\delta \rangle = \langle q_1^2 \rangle \frac{4\pi(\varepsilon' + 2)^2}{27\varepsilon'} \frac{f}{mC_t^3} g(f/f_0)$$

$$g(x) = 1 - (1 + x^2)^{-2} \tag{4}$$

$$f_0 = C_t / 2\pi l$$

where < > denote an ensemble average, q_1 is the deviation from the average charge, ε' is the real part of dielectric constant, m is the average mass of ions, C_t is the velocity of transverse sound, f is the driving frequency and l is the correlation distance which characterizes the strength of correlation (Smaller l corresponds to stronger correlation). Here it is found the dielectric loss tangent strongly depends on the spatial correlation between the charge deviations.

First, we consider the weak correlation limit which corresponds to the case of perfectly disordered charge deviation. In this case, assume f » $f_0 = C_t / 2\pi l$, thus $g(f / f_0) \fallingdotseq 1$ is derived. Finally, we obtain the following relation.

$$<\tan\delta>_w \propto \frac{f}{C_t^3} \tag{5}$$

In this case, it is found that the loss tangent is proportional to the frequency.

Next, we consider the strong correlation limit, where, for example, we can assume the charge neutrality is maintained within each until cell. In this case, assume f « f_0, thus substitute a lattice constant d into the correlation length l , then $g(f / f_0) \fallingdotseq 2(f / f_0)^2$ is approximately derived. Finally we obtain the following expression.

$$<\tan\delta>_s \propto \frac{d^2 f^3}{C_t^5} \tag{6}$$

In this case, the loss tangent is proportional to the cube of the frequency.

Comparing $<\tan\delta>_s$ of expression (6) with $<\tan\delta>_w$ of expression (5), estimate the ratio of $<\tan\delta>_s$ to $<\tan\delta>_w$. The ratio is equal to d^2f^2/C_t^2. Roughly estimate this factor, assuming $d \fallingdotseq 10^{-8}$ cm, $f \fallingdotseq 10^{10}$Hz, $C_t \fallingdotseq 10^5$ cm/sec, the order of the ratio is 10^{-6}. Thus, $<\tan\delta>_s$ for the case of correlation limit (that means highly ordered) is drastically reduced.

In the actual crystals, the case of intermediate correlation between the above two extremes would be expected to occur.

The above theory was introduced to explain the drastic change of the dielectric loss tangent in crystals with the inverse spinel structure and the 1 : 2 stoichiometric complex perovskites. Those crystals have equivalent sites which are occupied by different valent ions and the dielectric loss tangent at microwave frequencies is greatly influenced by the degree of order of cations with different valence charges. The $(Zr,Sn)TiO_4$ material has also the phase transition of $ZrTiO_4$ around 1130℃ from α-PbO_2 type high temperature structure with a random distribution of Zr and Ti ion to the low temperature structure with ordered phase of cations.

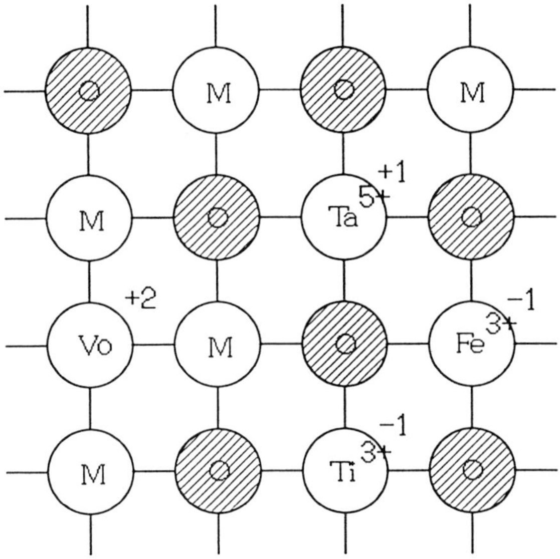

Fig.6. Simplified defect model for charge deviation of substitution of Ta^{5+} and Fe^{3+} for tetravalent cations and of oxygen vacancy. M and O denote tetravalent cation and oxygen, respectively

But in this case, as all kinds of cations have the same valence number +4, the phase transition or the degree of order of cations may have less influence on the dielectric loss tangent at microwave frequencies. This prediction also agrees well with the experimental results. When dopants with different valence number are introduced into the mother crystals and then we expect the influence of such defects with the charge deviations on the dielectric loss tangent, though the effects of such impurities are not considered to be so drastic as the order-disorder of ions.

Fig.6 shows the simplified defect model of $(Zr,Sn)TiO_4$ doped with donor and acceptor ions. M and O denote tetravalent cations and oxygens, respectively. The charge deviation of substitution of Ta^{5+} and Fe^{3+} for tetravalent cations are +1 and −1, respectively, and the charge deviation of oxygen vacancy is +2. These defects with charge deviations could be accompanied by the other kind of defects with different valence charge to maintain the charge neutrality in the crystal. The lattice defects with those deviated charges also increase the microwave dielectric loss.

Quantitative Analysis of Dopants and Oxygen Vacancies

The concentration of Ta and Fe ions in the grains was analyzed by the energy dispersive X–ray spectrometer for specimen B (0.47mol% $Ta_{5/2}$) and D (1.31mol% $Fe_{3/2}$).[11] As the result, the Tantalum spectrum was detected from the grains and not detected from the triple points. By comparing the spectrum intensity of Ta ions with that of Zr ions, the concentration of Ta ions was analyzed to be 0.9mol% as the average value in several grains. This value is much larger than the doped concentration. This error is considered to be caused by the analytical limitation of instruments. As a result, the Ta ions are estimated to concentrate in grains as much as the doped 4700 ppm because the spectrum was not detected from the triple points.

While Fe spectrum was detected both from the grains and the triple points in specimen D. The concentration of Fe ions in grains was analyzed to be 4.3mol% which is again much larger than the doped 1.3mol% concentration. Although an accurate value was not determined, a large percentage of doped Fe ions could be assumed to concentrate in grains. In specimen C, oxygen vacancies were generated by the annealing in reduced atmosphere in addition to Ta ions. The concentration of the oxygen vacancies was estimated to be 770ppm from the measured 0.0062wt% of reduction in weight before and after the annealing.

Dielectric loss

Pure $(Zr,Sn)TiO_4$ material like specimen A would be considered to have the ideal crystal structure. But as generally observed in many titanate ceramics, some fraction of oxygen vacancies are created in the crystal and some of Ti ions transfer from tetra to tri–valence to maintain the charge neutrality in the crystal. Thus the specimen A has high insulation resistance of 10^{13} $\Omega \cdot cm$ and the low $\tan\delta$ through the entire range of frequency.

It is surprising that tan δ of specimen B and C is as low as that of specimen A in spite of their large amount of Ta ions (4700ppm at most), and the additional oxygen vacancies of 770ppm for specimen C. According to the theory mentioned above, if we can assume the deviation of the charge of Ta ions are small and Ta ions behave equivalently like tetravalent ions, the reason could be understood. The Ta ions in $(Zr,Sn)TiO_4$ crystal could be Ta^{4+} if one of the electrons in 5d orbital of tantalum atom is bound and localized around the ion. We could also assume another mechanism. If the pentavalent Ta ions and trivalent Ti ions compensate the charge deviations each other with strong spatial correlation, we could also interpret the experimental results. The high insulation resistance, $10^{13}\,\Omega\cdot$cm, of specimen B could be also considered to be given by compensation of the holes and electrons generated from $Ta^{5+}$and Ti^{3+} ions.

In specimen C, oxygen vacancies were additionally generated by the annealing in reduced atmosphere. As the substitutional Ta ions and oxygen vacancies both generate free electrons, specimen C has the lowest insulation resistance of $10^8\ \Omega\cdot$cm and this causes the space charge polarization and increases K and tan δ seriously at KHz frequency region. The contribution of this space charge polarization seems negligible at microwave frequency region.

Compared to Ta ions, a much greater amount of Fe ions is expected to diffuse into grains, because the trivalent cations can easily substitute with tetravalent cations by creating oxygen vacancies to maintain charge neutrality. So the specimen D contains large amount of Fe ions and oxygen vacancies in the grains. If all of the 0.5wt% Fe_2O_3 had diffused into grains, the amount of substitutional Fe ions in the crystal would be 13000 ppm. Then the oxygen vacancies could have generated with half of that number at most to maintain charge neutrality. Thus, the drastic increase of tan δ at microwave frequencies of specimen D could be explained by the greater amount of defects with charge deviations, 13000ppm Fe^{3+} ions and 6500ppm oxygen vacancies at most while the electrons and holes are generated from substitutional Fe^{3+} ions and oxygen vacancies. Moderate lowering of insulation resistance in specimen D shows that the compensation of Fe^{3+} ions and oxygen vacancies are not completed.

CONCLUSION

This paper described the effects of dopants and oxygen vacancies on dielectric loss tangent of (Zr,Sn)TiO_4 ceramics. By the doping of Fe_2O_3, the microwave dielectric loss tangent increased drastically due to the charge deviation of a great concentration of Fe^{3+} ions and oxygen vacancies in the grains. When the specimen was doped with Ta_2O_5, then annealed in reduced atmosphere, its insulation resistance decreased down to $10^8\Omega\cdot$cm. The decreased resistance increased tan δ tremendously at low frequencies but slightly at microwave frequencies. The low dielectric loss tangent at microwave frequencies of Ta doped specimens would be explained reasonably if the Ta ions in (Zr,Sn)TiO_4 crystal could behave effectively like tetravalent ions or Ta^{5+} ions and Ti^{3+} ions could compensate each other with strong spatial correlation. But the above assumptions are needed to be proved in a future work to confirm the discussion.

REFERENCES

1. K.Wakino, K.Minai, and H.Tamura, "Microwave Characteristics of (Zr,Sn)TiO_4 and BaO– PbO–Nd_2O_3–TiO_2 Dielectric Resonator," J.Am.Ceram.Soc., **67** [4] 278–81 (1984)
2. R.E.Newnham, "Crystal Structure of $ZrTiO_4$," J.Am.Ceram.Soc., **50** [4] 216 (1967)
3. G.Wolfram and H.E.Göbel, "Existence Range, Structural and Dielectric Properties of $Zr_xTi_ySn_zO_4$ Ceramics (x+y+z=4)," Mat.Res.Bull., **16** [11] 1455–63 (1981)
4. A.E.McHale and R.S.Roth, "Investigation of the Phase Transition in $ZrTiO_4$ and $ZrTiO_4$–SnO_2 Solid Solution," J.Am.Ceram.Soc., **66** [2] C–18–C–20 (1983)
5. H.Ikawa, H.Shimojima, K.Urabe, T.Yamada, S.Udagawa, "Polymorphism in $ZrTiO_4$," Science of Ceramics **14**, 769–74 (1988)
6. R.Christoffersen and P.K.Davis, "Structure of Commensurate and Incommensurate Ordered Phase in the system $ZrTiO_4$– $Zr_5Ti_7O_{24}$," J.Am.Ceram.Soc., **75** [3] 563–69 (1992)
7. E.Schlömann, "Dielectric Losses in Ionic Crystals with Disordered Charge Distributions," Phys.Rev., **135** [2A] A413–19 (1964)
8. D.A.Sagala and S.Nambu, "Microscopic Calculation of Dielectric Loss at Microwave Frequencies for Complex Perovskite $Ba(Zn_{1/3}Ta_{2/3})O_3$," J.Am.Ceram.Soc., **75** [9] 2573–75 (1992)
9. Y.Kobayashi and M.Katoh, "Microwave Measurement of Dielectric Properties of

Low–Loss Materials by the Dielectric Resonator Method," IEEE Trans. on MTT, **MTT–33** 586–92(1985)
10. D.M.Iddles, A.J.Bell and A.J.Moulsen, "Relationships between dopants, microstructure and the microwave properties of ZrO_2–TiO_2–SnO_2 ceramics," J.Mater.Sci., **27**[23] 6303–10(1992)
11. N.Michiura, T.Tatekawa, Y.Higuchi and H.tamura, "Role of donor and acceptor ions on dielectric loss tangent of $(Zr_{0.8}Sn_{0.2})TiO_4$ dielectric resonator material," J.Am.Ceram.Soc., to be published

THE DEVELOPMENT OF A MULTILAYER CERAMIC ANTENNA SWITCH FOR WIRELESS COMMUNICATIONS

Rong-Fong Huang and Richard S. Kommrusch
Ceramic Technology Research Laboratory, Motorola Inc.
Albuquerque, NM 87109

ABSTRACT

Low fired multilayer ceramics (MLC) are useful in the manufacture of a broad range of high frequency components. When several basic components, such as, capacitors, inductors and transmission lines are integrated into a single multilayer ceramic package, a smaller and more consistent circuit can be achieved. An antenna switch was constructed using MLC. Aspects of the devices construction and electrical performance are presented.

I. Introduction

Multilayer ceramic technology has been widely used for the fabrication of substrates for high density IC packaging and surface mount components such as chip capacitors and inductors for electronic applications. Because of the ability to provide three dimensional integration, these multilayer devices can achieve the same or higher functionality with smaller size as compared with other processing technologies [1]. Recently, fueled by the continuing drive to reduce the size and weight of communication products, there has been a considerable interest to develop components or modules for high frequency applications using multilayer integrated technology.

To be useful for radio applications, the signal loss of these devices needs to be kept as low as possible (i.e. high Q) in the radio frequencies (RF) region which ranging from about 50 MHz to 3 GHz. The losses in these devices are mainly caused by the metal loss and dielectric material loss. But for most RF multilayer modules, the metal Q is the limiting factor that decides the overall Q due to the small size of the device. Therefore, high conductivity metals such as copper or silver have been exclusively used to make the practical RF devices. The low firing (≈ 850°C) dielectric tapes with dielectric Q in the range of 500 to 2000 have been found to be suitable for this application [2,3]. Many high frequency components and circuits such as couplers [4,5], voltage controlled oscillators [6], filters [7,8], and microwave modules [9] have been constructed using multilayer integrated technology.

In this paper, the design and fabrication of an antenna switch with harmonic filter module using low firing ceramic tapes with silver metallization is reported. The antenna switch can be used in cellular phones to switch the antenna between the transmitter and the receiver thus preventing signal power loss and providing receiver protection The switches harmonic filter prevents radiation of harmonic frequencies from the radio's transmitter and protects the receiver from possible interfering signals caused by products in the receivers mixer and the transmitters harmonics.

II. Experimental Procedure

The T/R switch with harmonic filter module was first designed and modeled. Based on the circuit design, the multilayer layout was carried out. In this module, both three dimensional helical and two dimensional zigzag transmission lines were designed in order to save the space occupied by the transmission lines. The multilayer layout design found a use for otherwise stray capacitance with a unique design that combined two capacitors into one single column to save space.

Conventional multilayer ceramic processing techniques similar to that seen in the fabrication of multilayer substrates for IC packaging were used for the construction of the module. Our processing equipment has a usable area of 2 by 2 inches this allowed a 4 by 4 array of modules to be processed at one time. Each layer of the array starts with a piece of DuPont green tape that is 4.5 mils thick before firing and 3.8 mils thick after firing. It has vias punched in it, and then, a cofirable silver ink was used to fill the vias and print a pattern on the layer. Ten mil vias were used and the transmission lines were typically 10 mils wide. The capacitors were made by stacking their plates and connecting the plates on alternate layers with 10 mil vias. After all twenty layers were printed, they were stacked together and laminated isostatically to form a single substrate. The green substrate was then scored so that individual units can be easily singularized after firing. A typical firing profile consists of two basic steps, binder burnout and densification. Starting at room temperature; 1) ramp up to 450°C in 45 min., hold at 450°C for 60 min. for binder burnout, 2) continue ramp to 850 °C in 85 min., hold for 20 min. for densification, and 3) ramp down to 600°C in 48 min. and then allow the substrate to cool naturally to room temperature. No special atmosphere was needed during firing. After firing, other parts, if needed, can be attached to the top of the substrate and then individual modules can be singulated by snapping the substrate array.

Since the size of the module as well as its connecting pads are small, a special test fixture was constructed using a printed circuit board (PCB). Using this fixture, the performance of the switch was measured. The PCB fixture exhibits 0.15 dB of loss which was deducted from the measurement values. A network analyzer (HP 8720C) and a current source set between 0 and 10 mAmp were used to test the device. To measure the frequency response between the transmitter port and the antenna port, the appropriate connectors on the test fixture were connected to the network analyzer and the unused receiver port was terminated in 50 ohms. The two other paths Rx to Ant. and Tx to Rx were then measured by reconnecting the network analyzer and load to the appropriate ports.

III. Results and Discussion

(1) Metallization

As mentioned earlier, the Q of cofired metallization is the dominating factor that determines the final Q of the transmission line. To achieve high Q at low firing temperatures, metal systems such as gold, silver, and copper pastes have been used. Silver and copper are the preferred metallization pastes for high frequency devices because they have lower losses and cost less than gold. However, the metal Q of conventional silver paste tends to be low. When tested using a stripline configuration, the commercially available silver ink such as DuPont6242D only produced a metal Q in the range of about 65% of the theoretical value. One method to improve the silver metal Q reported was to fire the device above the melting point of silver [10]. Because of the melting, the silver transmission lines has low defects, high density, and high Q.

In our development of MLC devices, a study was carried out to search for the most suitable cofirable thick film inks and processing methods for the fabrication of low loss transmission lines. Results showed that the metal Q of the transmission lines can be improved to reach 94% of the theoretical value using the conventional processing technique when a well formulated, silver paste was used. At the moment, it is not clear as to what causes the higher Q for some inks. However, the line's cross section and the ink's composition are two important parameters.

(2) Size Reduction Consideration for Embedded Transmission lines

Since the transmission lines occupied a large area in the switch module and its Q value is affecting the total insertion loss (I.L.) of the device, the performance of different multilayer transmission line designs was studied. The transmission line is traditionally designed as a two dimensional (2D) straight or zigzag line between two ground planes separated by a distance **b**. A three dimensional (3D) helical type transmission line was also designed, prepared and tested to see if the planar size of the line can be reduced. The 3D transmission line was formed by printing left and right hand semi-circles on alternate layers with vias connecting the semi-circles to form a continuous helical transmission line. Measured results on helical lines were compared with the 2D straight and zigzag striplines. The comparisons were made for a given ground plane separation **b** and line width **w**. The Q of the helical line was highest and had the smallest planer area. Zig-zag lines took less area than the straight lines but, their Q started to drop as their area was decreased. In some cases a helical line would have 70 to 80 % more Q than a zig-zag line of the same area. It was also noted that for a given line width **w** the characteristic impedance (Zo) of the helical line was higher than that of a straight line and the Zo of the zig-zag line was lower than than that of a straight line. In some applications the ability to make high Q lines with higher Zo's is an advantage.

(3) MLC T/R Switch with harmonic Filter

The switch is designed to operate from 890 to 915 MHz for the transmitting band with bias applied and 935 to 960 MHz for the receiving band with no bias. The I.L. in the Tx to antenna mode is a critical parameter for the device, which needs to be kept as low as possible. One type of antenna switching circuit that has found use for this application consists of 2 transmission lines, 2 PIN diodes, and 3 DC blocking capacitors. Figure 1 shows the basic circuit schematic for this switch. The loss is determined by the quality of the components. Once a low loss PIN diode is picked the Q of the transmission lines become the limiting factor. The 3 blocking capacitors keep the desired DC control signal out of the rest of the radio and also keep other DC signals from the radio off the switch. This type of T/R switch has been made by others and typically achieves an I.L.'s of less than 1 dB in both the Tx and Rx to Antenna paths.

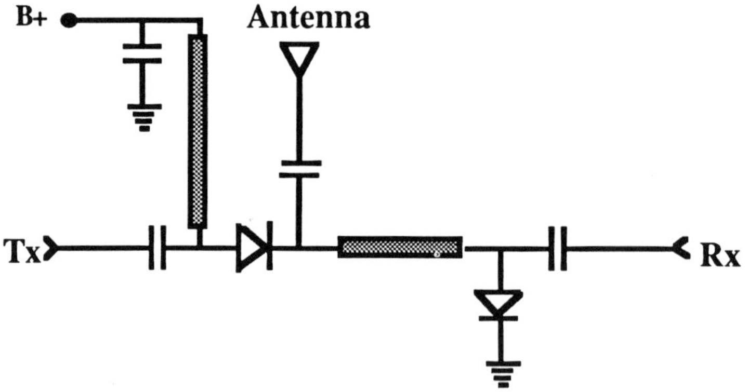

Figure 1: Conventional T/R switch schematic.

The preferred type of transmission line for this application is stripline because its two ground planes reduce the chance of interference by keeping its electrical signals from coupling to other parts of the radio. A microstrip or an unshielded inductor design can suffer from undesired electrical coupling. The multilayer capacitors are normally made by stacking plates of a given area and number to create the desired capacitance according to the formula C= ε_r K N A / t, where ε_r is the relative dielectric constant, K=8.854 x10^{-14} F/cm, N is the number of dielectric layers, A is the area of the electrode and t is the thickness of a layer. When a multilayer DC blocking capacitor is implemented between the ground planes of stripline transmission lines, undesired stray capacitors to ground are created. Figure 2 shows a DC blocking capacitor and its stray capacitance to ground.

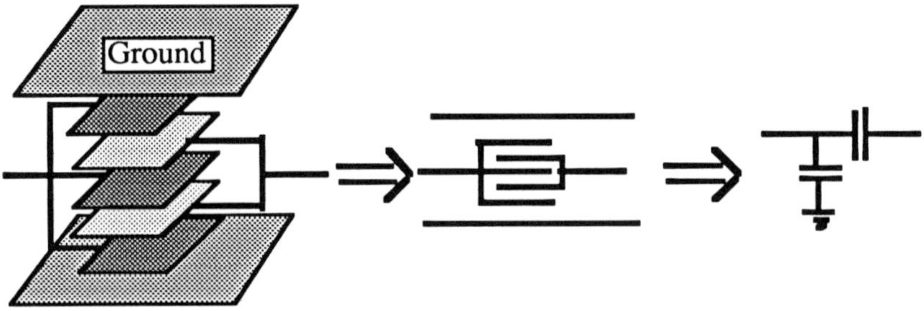

Figure 2 DC Blocking Capacitor With Stray Capacitance to Ground

Figure 3 shows where these undesired capacitors modify the original circuit shown in figure 1. The influence of the stray capacitors is big enough to change the performance of the switch significantly. Normally the stray capacitors cause a nuisance that needs to be kept as low as possible. Very often the design has to be changed to reduce the effect of stray capacitance.

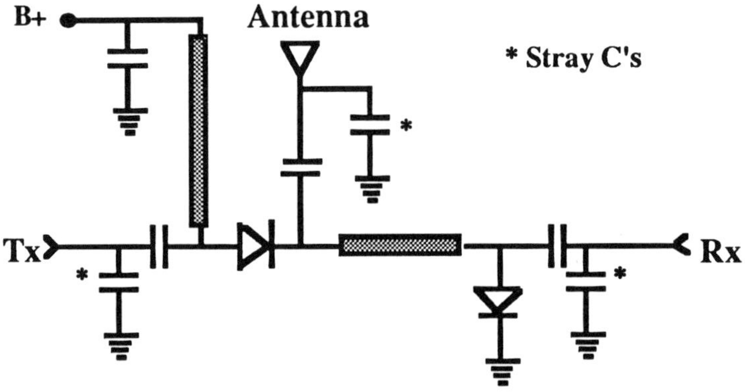

Figure 3: T/R switch with stray capacitors caused by multilayer structure.

An alternative approach was to find a use for the stray capacitors. When the application of the switch was reviewed, it was determined that most transmitters need harmonic filters and harmonic filters need capacitors to ground. Figure 4 shows a five pole harmonic filter made with three capacitors and two pieces of transmission line. A close examination of figures 3 and 4 revealed that the three stray capacitors shown in figure 3 can be used and only one additional capacitor was needed to make an T/R switch with a harmonic filter (TRH) in each branch.

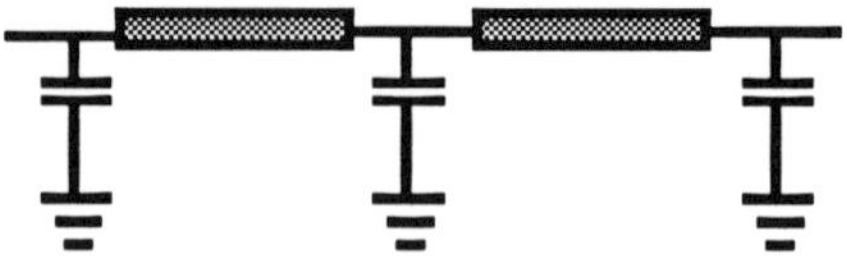

Figure 4: A five port harmonic filter.

The final design of the TRH is shown in figure 5. Some additional changes are included that improved TRH. One improvement was the use of a 3 turn helical transmission line for the line near the Tx port. Another improvement increased the isolation between the Tx and Rx ports by moving the DC bias port. In the original design the small inductance of the PIN diode in the on state was limiting the isolation between the Tx and RX ports to 13 dB. This was improved to 25 dB by moving the B+ port and the capacitor shown in figure 3 that provided an RF ground to the quarter-wave line near the Tx port to near the RX port as shown in figure 5. The value of the capacitor was reduced so that it would series resonate with the diodes inductance and provide a good RF ground. After the circuit was re-optimized on a computer, it was found that the three lines use in the harmonic filter had a total length of the quarter -wave line they replaced, thus no additional module area was required for them. When the final design was built in multilayer it was found to be a little smaller than the design without the harmonic filter.

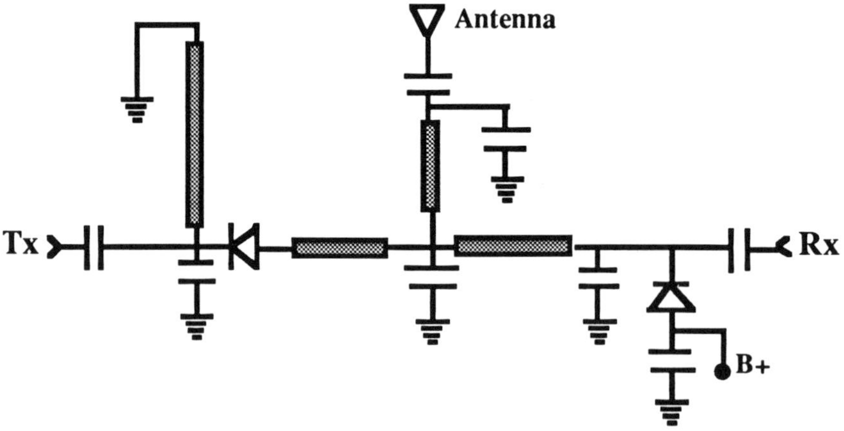

Figure 5: The circuit layout of the T/R switch with harmonic filter.

The final module, which contains four transmission lines and eight capacitors, was 360 mils x 300 mils x 72 mils in size before the diodes were attached to the top. Figure 6 shows the cross-section of the module.

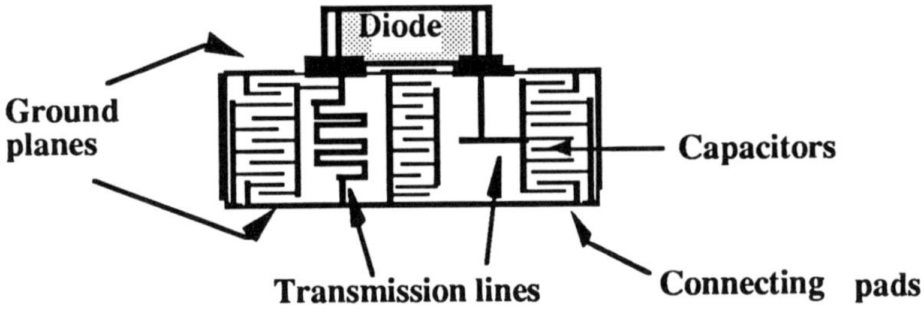

Figure 6: The cross section of the T/R switch with harmonic filter.

(4) Electrical performance of the TRH switch

The frequency response of this module is shown in Figure 7, with the Motorola diodes (MMBV3401L) attached. Typical performance of the switch is shown in Table I. The insertion loss of Tx to antenna and Rx to antenna are less than 1 dB with 0.75 dB typical. From figure 7, it can be seen that the switch has a relatively wide band response. The Tx to antenna and the Rx to antenna have similar responses with I.L.'s under 1 dB. The response of the integrated harmonic filter can be seen in the first two curves by its 3 dB cutoff frequency between 1.2 and 1.3 GHz. The module was further tested up to 6 GHz to determine the rejection on high order harmonic frequencies. Results showed that the isolation on other harmonic frequencies were all greater than 20 dB.

Table I: Performance of the TRH Sample.

	Tx band (890 to 915 MHz)	Rx band (935 to 960 MHz)	1.8 GHz Attenuation
Tx to Antenna			
Tx Insertion Loss (Bias on)	0.72-0.74 dB @ 5 mA		22 dB
Rx isolation (Bias off)		19 dB	16 dB
Rx to Antenna			
Rx Insertion Loss (off)		0.7 dB	26 dB
Tx Isolation (on)	28 dB@ 5 mA		14 dB
Tx to Rx isolation			
Bias on	40 dB		20 dB
Bias off		19 dB	15 dB

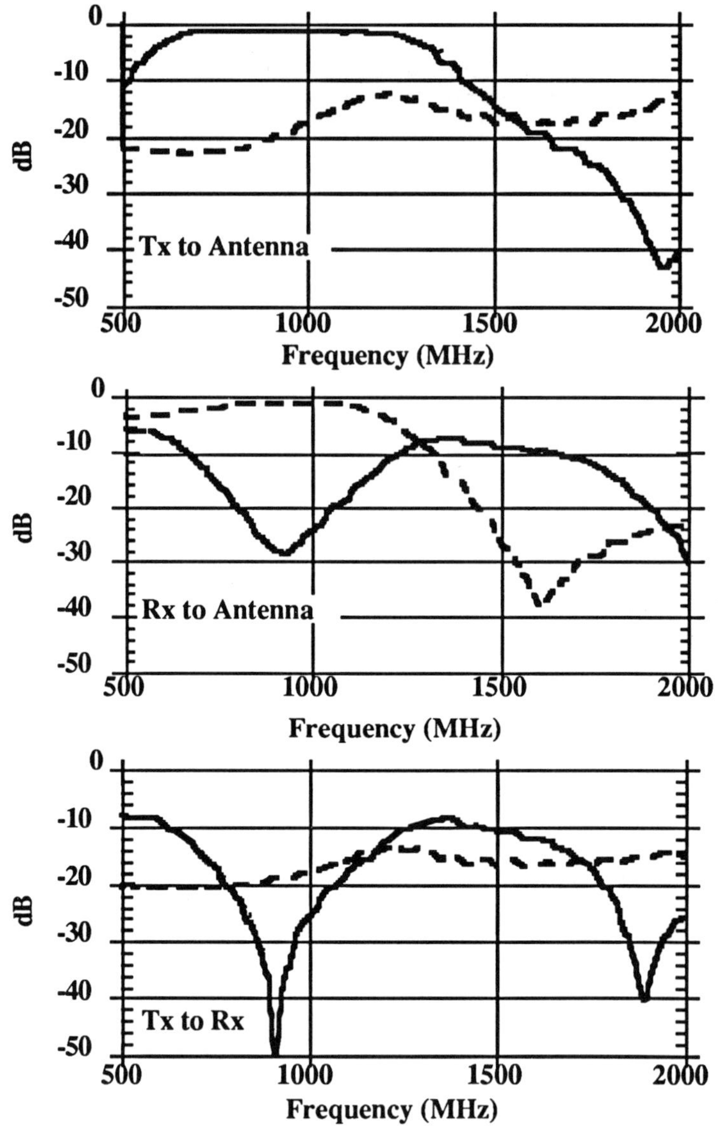

Figure 7: The frequency response of T/R switch module with integrated harmonic filter. The solid lines are bias on and the dotted lines are bias off responses, respectively.

IV. Conclusions

Multilayer ceramics using low fired dielectric ceramic tapes and silver metallization are suitable to build low loss integrated RF circuits for wireless communication. A ceramic antenna switch with integrated harmonic filtering function was developed. The antenna switching circuit contains eight capacitors and four transmission lines which are ideal to be built using the multilayer ceramic technology. The performance of the device was measured and showed to be suitable for high frequency applications. A harmonic filter was added to the module without increasing the antenna switch's size. The techniques to utilize stray capacitance associated with the integrated structure for other useful functions were demonstrated. With proper design, the multilayer integrated modules exhibit the advantages of size reduction, lower insertion loss, good RF shielding and a lower parts count for future radios.

References

1. A. Weinberg, "High Density Electronic Packaging Utilizing Vertical Integration and Low Temperature Cofired Ceramics," pp 618-625 in ISHM '90 Proceedings, Chicago, (1990).
2. R. L. Brown, P. W. Polinski, and A. S. Shaikh, "Manufacturing of Microwave Modules Using Low-Temperature Cofired Ceramics," pp1727-1730, 1994 IEEE MTT-S Digest, TH4C-6 (1994).
3. J. R. Tyler, and D. J. Gasper, "Evaluation and Characterization of Low-Temperature Cofired Ceramic Structures for Microwave Applications," pp390-399 in ISHM '89 Proceedings, Baltimore (1989).
4. D.J. Davis, Jr., and V. Insetta, "Miniature Monolithic Ceramic Coupler for Electronic Circuits" U.S. patent # 5,162,970; Nov (1992).
5. T.K.Kawasaki, S.Sakuragi, T. Ikeda, and T. Kuroda, "Multilayered Ceramic Type Electromagnetic Coupler Apparatus," U.S. patent # 5,214,392; May (1993).
6. Y. Yasukawa, S.-Y. Nakai, K. Kawamura, and M. Kobayashi, "A Miniaturized VCO Using Multilayer Ceramic Technology," pp 309-313 in IMC 1992 Proceedings, Yokohama (1992).
7. H. Hayashi, T. Ikeda, and S. Nishigaki, "Newly Developed Multilayered Small Size LC Chip-Elements for EMC Noise Filters Made by Low Temperature Fireable Ceramics," pp 508-512 in ISHM '91 Proceedings, Orlando, (1991).
8. T. Ishizaki and T. Uwano, "A Stepped Impedance Comb-Line Filter Fabricated by Using Ceramic Lamination Technique," pp 617-620, 1994 IEEE MTT-S Digest, WE1C-4 (1994).
9. E.L. Rich, III, "Multi-layer Single Substrate Microwave Transmit / Receive Module" U.S. patent # 4,967,201; Oct. (1990).
10. H. Shimoda, N. Ishitobi, and K. Kawamura, "Development of a High-Q Multilayer Ceramic Resonator," Jpn. J. Appl. Phys. Vol. 31, pp3160-3163 (1992).

INFLUENCE OF CHEMISTRY AND PROCESSING ON THE ELECTRICAL PROPERTIES OF $Ba_{6-3x}Ln_{8+2x}Ti_{18}O_{54}$ SOLID SOLUTIONS

T. Negas, Trans-Tech Inc., Adamstown, MD 21702
P. K. Davies, Univ. of Pennsylvania, Philadelphia, PA 19104

ABSTRACT

Contrary to most literature, ceramics used in many wireless communication circuits belong to a solid solution series, $Ba_{6-3x}Ln_{8+2x}Ti_{18}O_{54}$, where x depends on the lanthanide (Ln). The condition where temperature coefficient of resonant frequency (T*f*) is near zero ppm/°C is explored from structural and polarizability considerations using Sr^{2+}, Pb^{2+}, Bi^{3+}, and Ln^{3+} substitutions. Processing produces preferred orientations in ceramics that are used to estimate very large intrinsic electrical anisotropy.

I. INTRODUCTION

Ceramics based on Ba,Ln-titanates (n:m:p; $nBaO{\bullet}mLn_2O_3{\bullet}pTiO_2$; Ln = Sm,Nd) are designed into circuits for numerous wireless applications below 5 GHz. They have high dielectric constant (É) in the 80-90 range; modest Q (1.5-4K, ~3 GHz); and low temperature coefficient of resonant frequency (T*f*), of dielectric constant (Te), and of capacitance (Tc), all near zero ppm/°C. For later reference, these coefficients are related to a first approximation by (1)T*f*~ $-$Te/2$-\alpha$ and (2)Tc~ $-$2T*f*$-\alpha$ where α is the linear thermal expansion coefficient. Ceramics are used mostly as metallized, TEM mode, coaxial filters and inductors. Less frequently, they serve as unmetallized TE_{01} resonators or as substrate platforms for thin/thick film circuits. High É reduces circuit/device size, reasonable Q allows distribution of signal with minimal loss, and low temperature coefficients avoid drift in frequency due to temperature variations. Although many factors can cause deviations, T*f* and É ideally should be independent of applied field, e.g. TE, TEM, capacitance (Ci), etc.

Despite industrial development and market acceptance of these ceramics, the open literature has been of limited help. It is characterized by considerable confusion and contradiction and is often misleading. After the first exploratory work by Bolton (1), researchers, often not cognizant of related works, have promulgated evidence that the phase is 1:1:5 (2-6), 15:19:72 (7), 1:1:4 (8,9), 1:1:4.25 (10), and 2:2.5:9 (11). Recently, the first three are accepted as "compounds" (12).

Matveeva et al.(7) determined the only widely available crystal structure. Using a Pr composition, they showed that it is orthorhombic, a~12, b~22, and c~3.8Å(x2), with the space group Pba2 (3.8Å subcell) and a chemistry of 15:19:72. The ab plane contains strings of corner-sharing Ti^{4+}-octahedra that frame edge-shared polyhedra described as pentagonal columns (A) and distorted eight coordinated sites (A'). Ti^{4+}-octahedra corner-share along c but are counter-tilted off parallelism, thereby doubling this dimension. Thus, the structure is a hybrid of perovskite and tetragonal bronze elements. It formulates as $A_4A'_{10}Ti_{18}O_{54}$

(1/2 cell) but, interestingly, Ba^{2+} partly occupies A. The Ln^{3+} and the balance of Ba^{2+} are disordered in A'.

In unpublished work, Beech et al. (13) determined that the phase is not a point compound but belongs to a solid solution (ss) series. Their structure (Pbam, 1/2 cell) is similar to the topology in (7) but site occupancy differs. They determined the structural formula as $A_4A'_8A''_2Ti_{18}O_{54}$ wherein Ba^{2+} and Ln^{3+} completely fill A and A' sites, respectively. Excess Ba^{2+} and Ln^{3+} occupy A''. Thus, they advanced the formalism for a ss, $Ba_4Ln_8(Ba_{2-3x}Ln_{2x}V_x)Ti_{18}O_{54}$ or $Ba_{6-3x}Ln_{8+2x}V_xTi_{18}O_{54}$, where V = vacancy. When A'' is filled completely with Ba^{2+}, the endmember is x = 0(3:2:9). When A'' lacks Ba^{2+}, the terminal ss is x = 2/3(2:2 1/3:9). Accordingly, this range eliminates all prior formulations, including 15:19:72 (x = 0.75), except 1:1:4 (x = 0.5). Such structural and ss details should be confirmed for Ln larger than Nd to prove that pentagonal A sites contain Ba^{2+} exclusively.

Electrical properties of high É phases continue to be puzzling lacking an accepted foundation of crystal chemistry, structure, and phase equilibrium. Kolar et al.(14) proposed that the −Te(+T*f*) of the 1:1:5Nd "compound" and of the 1:1:4Nd "bulk composition" are tuned closer to zero in ceramic composites formulated to contain $Bi_4Ti_3O_{12}$ (and/or "$Bi_2Ti_2O_7$") which has +Te(−T*f*). Wakino et al. (15) accepted 1:1:5Nd but showed that Pb^{2+} lowers the T*f* of the 1:1:4Nd composition to zero. In (4-6), researchers accepted 1:1:5Sm but had great difficulty explaining why ceramics had a −T*f* that seemed to be influenced by "metastable phases" such as $Ba_2Ti_9O_{20}(B_2T_9)$, $BaTi_4O_9(BT_4)$, and TiO_2. Interestingly, these phases are reported as "impurities" for the standard X-ray diffraction (XRD) pattern of 1:1:5Nd (16). Finally, all pertinent work has ignored Clausius-Mosotti (C-M) theory which relates É, molar volume (V_m), and net dielectric polarizability ($A_D = \Sigma a_D$ of individual ions) according to (3)$É = 3V_m + 8\pi A_D/3V_m - 4\pi A_D$ (17). Differentiation and evaluation of eq. 3 specify how V_m and A_D must be varied to obtain a desired change in É. Differentiating with respect to temperature yields Te(T*f*,Tc) plus expansion and TA_D coefficients. The first operation provides guidance in choosing structures/ chemistries for study, as well as in selecting appropriate substitutions in concert with knowledge of ionic radii and V_m (e.g., via XRD). Shannon (17) has developed reasonable values of a_D. Evaluation of coefficients from the latter, however, suffers due to lack of knowledge of Ta_D, TA_D terms which typically are assumed to be small, isotropic, and constant (18). The problem is far more complicated since É and A_D are tensors.

This report highlights linkages among basic chemistry/structure, processing, and properties using several examples. It is intended to provoke continued work on these enigmatic materials. Extensive data and analyses will become available in the future (19). Many abbreviations throughout the text are summarized for rapid reference in the Appendix. Reported V_m does not include doubled c̲. Rather than ionic radius, some measure of A_D as a scalar is adopted to relate to bulk chemistry and to C-M.

II. RESULTS AND DISCUSSION

A. General Chemistry and Properties

Figure 1 illustrates properties of 1:1:4 (x = 0.5) variants. From La^{3+} through Eu^{3+}, V_m and A_D decrease such that É decreases continuously in accord with C-M. Simultaneously, Q̲ increases and T*f* transitions from positive to negative at fixed

A_D and V_m close to x = 0.5Sm. Discussion of Q is reserved. Ceramics based on La and on Eu are not used because T*f* of the former is far too positive while the latter are too expensive. Thus, practical ceramics are based on Sm and Nd variants. It will become obvious that Fig. 1 highly oversimplifies the ss, but, for imprecise reasons, it correctly predicts that T*f*~0 should be obtained by substituting larger Ln^{3+} for Sm^{3+} at x = 0.5. Approximate chemistries can be calculated using the predicted A_D, then slightly adjusted, if necessary, as demonstrated in Table I. Within error, all cluster about similar V_m and A_D and, therefore, a nearly constant É. This principle seems to hold almost quantitatively for variants containing mixed Ln^{3+} but it fails to account for property variations when other cations are substituted.

Figures 2 and 3 summarize data for Sm and Nd compositions. From T*f* and XRD, limits for each ss series can be approximated. T*f*, rather than É or even Q, is preferred since it is most dependent on phase content and is far less sensitive to ceramic porosity (19). Complete tabulations of relevant data are found in (19). Both series terminate near x = 0.7 which, within error, supports Beech et al. At x = 0.75 (15:19:72), ceramics and equilibrated powders are of multiple phase. Both series, however, do not originate at a common x near zero nor at x = 0. This disagrees with Beech et al. in fact rather than in principle. The discrepancy could be due to the use of "sol-gel" syntheses in (13) rather than conventional methods. Interestingly, Q maximizes in the range x = 0.5-0.7 where the concentration of cation vacancies maximizes. Thus, useful ceramics are confined to x > 0.4 and the

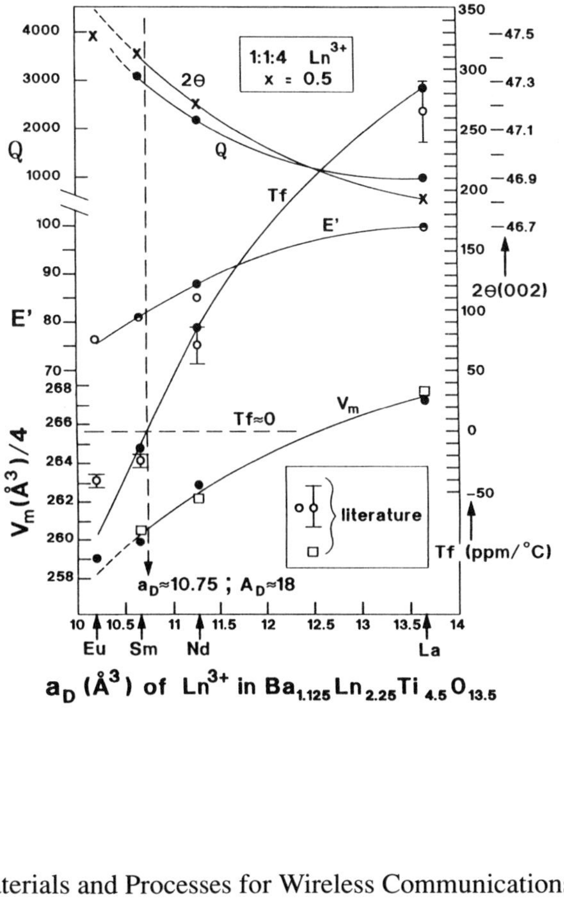

Fig. 1
Electrical (~3GHz), XRD (CuKα) properties of x = 0.5Ln phases. a_D = $\Sigma a_D Ln^{3+}/4$; A_D = $\Sigma a_D Ln^{3+}$ + $\Sigma a_D Ba^{2+}/4$.

TABLE I
Electrical Properties of Selected $Ba_{6-3x}Ln_{8+2x-y}Bi_yTi_{18}O_{54}$ with Tf_r Near Zero

Composition	$\acute{E}_r$	Q	Tf_r(ppm/°C)	A_D($Å^3$)	V_m($Å^3$)
x=0.5					
$Sm_{7.65}Nd_{1.35}$	83.6	3230	-3	17.96	1041.04(.4)
Sm_5Nd_3Y	81.2	2700	1	17.84	1041.50(.6)
$Sm_{8.55}La_{0.45}$	82.5	2850	-1	18.01	1040.60(.3)
$Sm_{8.1}La_{0.9}$	82.5	2700	7	18.16	–
$Sm_{7.2}La_{1.8}$	80*	2650	30	18.46	1046.50(.2)
x=0.4					
$Sm_{7.44}Nd_{1.36}$	86	2030	3	18.20	–
x=0.6					
$Sm_{7.3}Nd_{1.9}$	79.6	3320	3	17.75	–

All electricals ~ 3 GHz. $\acute{E}_r$ fluctuations influenced by fired density. *, lowest fired density (5.53 vs>5.60g/cc). Polarizability, $A_D = \Sigma\ a_D(Ba^{2+}) + \Sigma\ a_D(Ln^{3+})/4$. V_m=molar volume.

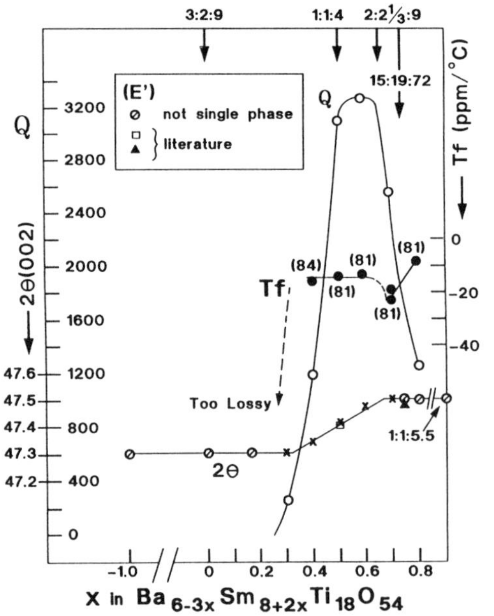

Fig. 2
Electrical (~3 GHz), XRD (CuKα) properties of Sm compositions delimiting the ss. 1:1:5.5 does not refer to a value of x, but contains terminal ss plus B2T9 and TiO2.

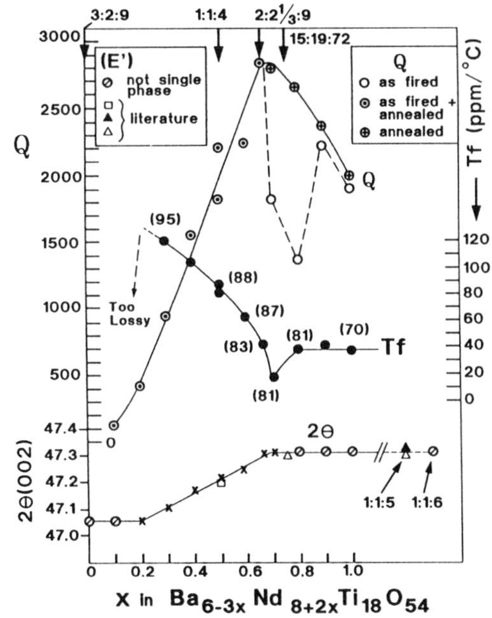

Fig. 3
Electrical (~3 GHz), XRD (CuKα) properties of Nd compositions delimiting the ss. 1:1:5 and 1:1:6 do not refer to a value of x, but contain B_2T_9 and TiO_2 plus a ss near terminal. Annealing and Q are beyond this scope.

question of ss stability toward x = 0 is academic. Figure 4 is a preliminary phase diagram for the Sm system. The Nd system is similar, but the extent of ss differs slightly (Fig. 3) and there is compatibility between ss and $Nd_4Ti_9O_{24}$(Nd_2T_9). This requires a primary field for Ln_2T_9 that also exists in the La system (19). Topology of the liquidus was determined for two reasons. First, the liquidus and subsolidus when in mutual conformance with Alkemade theory provide mutual validity. Second, reaction paths for bulk chemistries and phase contents can be followed so that intervention of stable and/or metastable melting of ceramics during processing can be predicted and, if necessary, altered or avoided (19). All diagrams prove that a ss which melts along a solvus, rather than a congruent point compound (3), must be present. Eu and La systems (19) are quite different. In fact, Euss is limited to x = 0.4-0.5 and it is not clear whether 1:1:4 (x = 0.5) actually is stable. The La system shows that x = 0 may be approached, but it is not clear whether the barrier at x = 2/3 is violated. Thus, from Eu through La the ss has different stability limits and details of the subsolidus/liquidus follow accordingly.

Figure 4 conflicts with (8) in which a tie-line from TiO_2 to a ss member is absent. This equilibrium, clearly precluded in the Eu and La systems (19), is demanded in both Sm and Nd systems. It explains B_2T_9/TiO_2 as "metastable" products (4-6) and as "impurities" (16) for 1:1:5 "compounds". Figure 4 also explains why 1:1:5Sm ceramics can show considerable T*f* variability. In (5,6), ceramics were fired, unknowingly, near and above peritectic melting with consequences detailed in (19). Microstructures (4-6) also show evidence for transitory melts. It has been demonstrated that the −T*f* of Smss can be offset to and through zero simply by producing ceramic composites containing $Sm_2Ti_2O_7$

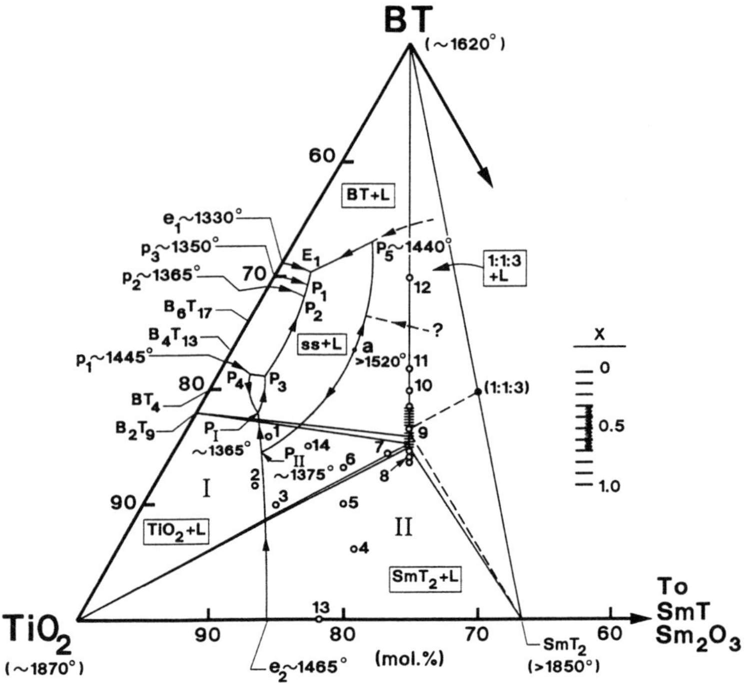

Fig. 4 TiO_2-rich portion of Sm system showing topology of liquidus, critical compatibilities, and ss (hatched). P_I-P_3-P_4 outline the primary field of B_2T_9.

and/or TiO_2 (20). Both secondary phases have a +T*f*. The +T*f* of Ndss can be offset toward zero with excess $Nd_2Ti_2O_7$ of −T*f*, implicit in (1,15), but É decreases precipitously.

Combining elements of Figs. 1-3 yields a more complicated picture of the ss, Fig. 5. T*f* and É decrease along an apparent V_m continuum manifested by c. Terminal Ndss transitions to the Smss series through a range of forbidden A_D and T*f*~0. Table I and Fig. 5 show that this region can be attained by appropriate Ln^{3+} substitutions. In fact, the data demonstrate that every Smss can be tuned into this region. Thus, the T*f*~0 state is unique and will be revisited after addressing a related complication.

B. Electrical Anisotropy, Processing, and Chemistry

Figure 6 summarizes our methodology in forming green compacts and in dissecting densified ceramics to obtain samples for electrical measurements. Orthogonal samples (A,B) contain forming as a unique vector which can be interrogated by perpendicular TE_{01}(GHz) and Ci(1MHz) fields. Type A samples are most important although all sample orientations are discussed and interrelated in (19). Silver electrodes were applied to samples for Ci measurements. Many tests, including confirming measurements using TM_{01}(normal to TE), were performed

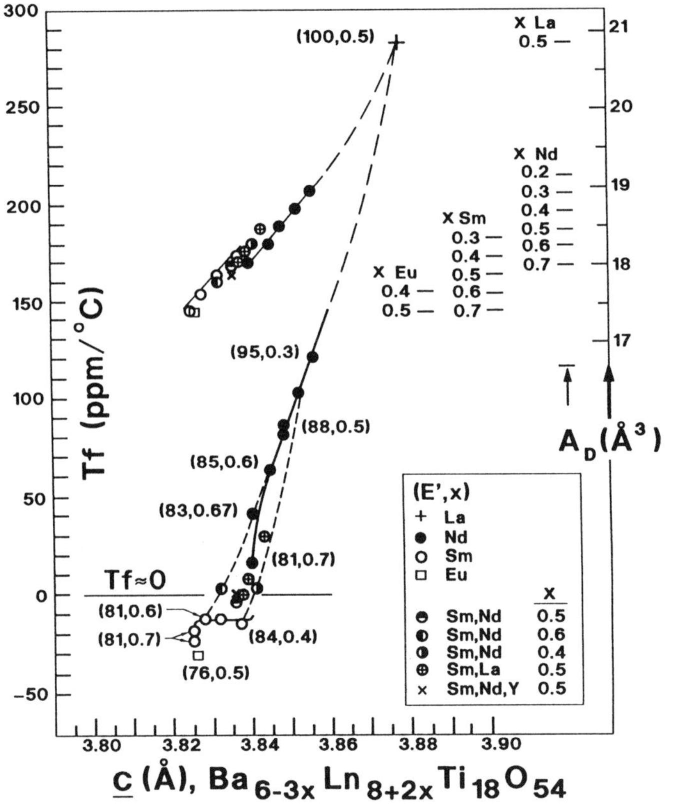

Fig. 5 Tf, net polarizability (A_D) vs. c-axis continuum of Lnss showing transition through $Tf \sim 0$. $A_D = \Sigma a_D Ln^{3+} + \Sigma a_D Ba^{2+}/4$ for every x. Examples of É shown. Smss and Ndss at equivalent x are connected by dashed lines.

before accepting the Ci field routinely. The notation Tf_N^M, $\acute{E}_N^M$, etc. is adopted where M is sample type and N is forming (I = isostatic, H = uniaxial hard die, E = extrusion). Tf values from Tc_N^M are computed via eq. 2 assuming α ~9ppm/°C.

The ss manifests many features proving that it is not an isotropic entity. Bonding is highly anisotropic and the cell is highly anisometric. The reciprocal cell predicts an equilibrium growth habit characterized as prisms or tablets elongated along c. Figure 7 suggests that this, indeed, is the case and other evidence will follow. Thermal expansion along c is 10-12 while a,b show 4-6 ppm/°C (19). Under proper circumstances (19), E ceramics show differential shrinkage of 4-8% while H yield 1-2% using identical powder. I ceramics never display shrinkage anisotropy.

Given many manifestations of anisotropy, electrical anisotropy (EA) is to be expected. One unique combination of É and Tf is not possible except under

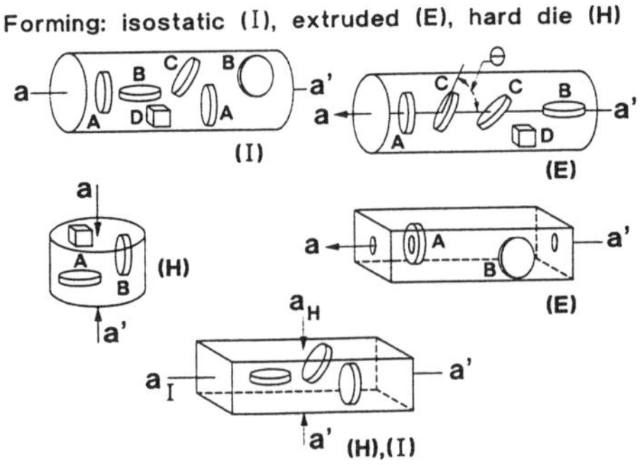

Fig. 6
Extracted A, B, etc. samples from formed, fired ceramics for electrical measurements. Samples are relative to aa′, the forming direction for H,E and the length for I.

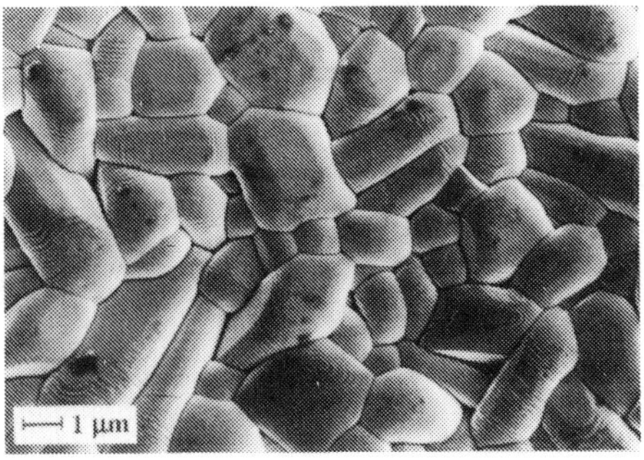

Fig. 7
Random section through elongated growth habit of typical ss.

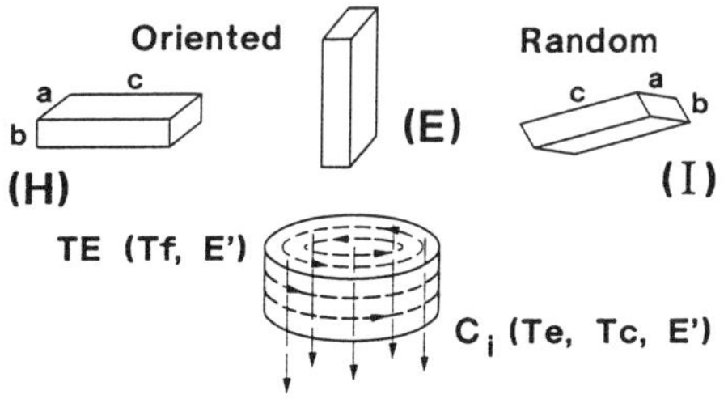

Fig. 8
Extracted type A sample (Fig. 6) relative to crystallite orientations imparted by forming (I, H, E). Sample can be interrogated by orthogonal fields.

special conditions. Intrinsic properties of the ss depend on axial direction, and ceramic processing can produce infinite variations for any fixed bulk chemistry. Before reviewing examples from the data, the relationship among an A sample (Fig. 6), its forming information, and the orientation of TE and Ci (or TM) fields with respect to a microstructural feature such as preferred crystallite alignment (PCA) must be understood. Figures 6 and 8 combine to show how oversimplified TE and Ci fields interrogate and extract information from a ceramic having cylindrical geometry. TE is azimuthal and congruent with the disc. Ci interrogates the volume normal to TE. Three idealized cases of PCA involving an elongated tablet (i.e., the ss) are illustrated. Even if PCA is mixed with random orientations, TE extracts T*f*,É containing information from ac and ab associated with H and E, respectively. Ci, however, provides Tc(T*f*,Te),É containing information from b and c, respectively. Both fields sense random alignments for I and, therefore, must yield similar properties.

Important conclusions from Figs. 6,8 and supporting electrical and XRD evidence follow:

a) c and b orient normal to and parallel with the uniaxial pressing direction of the H process, respectively; ac planes in type A samples are randomly disposed normal to pressing;

b) ab planes and c align normal to and parallel with the high shear direction of the E process, respectively; crystallites "wobble" and rotate about c along this direction; and

c) being isostatic, the I process does not impart PCA, and EA is absent; regardless of sample type and applied field, properties are invariant and define Tf_r,$É_r$ of randomly oriented crystallites; thus, $Tf_r;É_r = Tf_I^M;É_I^M$.
A finite concentration of the ss rather than a precursor phase is necessary in the starting powder before PCA can be initiated by forming (19). During densification, the amount of PCA is enhanced by mechanisms beyond this scope. Final ceramics always consist of a mixture of aligned and random crystallites. If powders lack ss, PCA and EA are absent regardless of forming and sample type. When PCA is present, measured properties depend on its nature and amount, sample type, orientation of the applied field, and intrinsic EA of the ss. In turn, $Tf_r;É_r \neq Tf_{H,E}^M;É_{H,E}^M \neq Tf;É$ (from $Tc_{H,E}^M$; Ci field).

Analyses of PCA are simplified using the Lotgering XRD method (21) which approximates volume fractions of aligned (V_l) and random ($1-V_l$) crystallites. Using simple mixing rules, intrinsic EA can be estimated from experimental Tf_N^M, $É_N^M$ and calculated V_l. For example, (4)$Tf_N^M \sim V_l Tf(\text{intrinsic}) + (1-V_l) Tf_r$ and (5)$\log É_N^M \sim V_l \log É\,(\text{intrinsic}) + (1-V_l) \log É_r$. Results yield sign and rough magnitude of intrinsic properties that are not accurate because of cumulative measurement errors, particularly in V_m when small. The necessary assumption of fixed V_l rather than a continuum of alignments (best determined from pole figures/ sophisticated XRD) probably leads to greatest error because Tf_N^M and $É_N^M$ also include information from less than perfect orientations. Nevertheless, ss variants reveal remarkable intrinsic properties summarized by:

a) $Tf(||\underline{b}) < Tf(\underline{a} \cup \underline{b})$ where both are always highly negative;
b) $Tf(||\underline{c})$ is always highly positive; $Tf(||\underline{a})$ cannot be isolated but must fall between $-Tf(||\underline{b})$ and $+Tf(||\underline{c})$ since $Tf(\underline{a} \cup \underline{c})$ is always $< Tf(||\underline{c})$; and
c) $É(||\underline{b})$, $É(||\underline{a}) << É(||\underline{c})$.

The data demonstrate that the direction of corner-sharing, c, has highly positive character and large É like perovskites, $CaTiO_3/SrTiO_3$. This could relate to recent work (22) suggesting that the ss might be ferroelectric. This possibility cannot be deduced from evidence in (23), but merits more work if the space group is not centered.

Smss and Ndss series have $-Tf_r$ and $+Tf_r$, respectively. Each, therefore, has different $Tf(\underline{a} \cup \underline{b} \cup \underline{c})$. In the former, $-Tf(\underline{a} \cup \underline{b})$ must overwhelm $+Tf(||\underline{c})$ while $+Tf(||\underline{c})$ must dominate the latter. Thus, the $Tf_r \sim 0$ state must be obtained when $Tf(||\underline{b})$ and $Tf(||\underline{a})$ provide enough negativity to cancel $+Tf(||\underline{c})$. Appropriately, all properties given to this point refer to Tf_r, $É_r$. Literature data represent Tf_H^A, $É_H^A$ or Tc_H^A, $É_H^A$ (from Ci) because the H process always is used to produce small tablets of type A orientation.

Eq. 1 relates frequency and É through temperature, thereby showing that Tf and Te depend on A_D. É,Tf,Te, and A_D, however, are tensors. Although eq. 3 is useful to help confirm general trends of $É_r$ from V_m and a scalar A_D, Tf_r cannot be predicted from fixed a_D values of ions (17). The problem is compounded since every ss contains Tf components of opposite sign and intrinsic EA is large. On the other hand, changes in A_D due to cation substitutions and to temperature are caused by perturbations in bonding that should manifest macroscopically. Thus, variations in a,b,c and V_m are adopted to interpret properties in phenomenological fashion, realizing that effects due to thermal expansion are not explicit. Some interesting aspects of bonding and thermal motion are reserved for summary after reviewing the data. The Beech et al. formalism illustrates the simplest cases with some complications. Each ss series contains a progressively smaller Ln^{3+} of smaller a_D.

Within each, smaller Ln^{3+} replaces Ba^{2+} of larger a_D, thereby creating vacancies. Both cases collapse V_m, causing Tf_r to decrease or remain nearly constant (Figs. 1,5). This correlation is never violated regardless of A_D variations from type and complexity of substitutions and of how a,b,c change to attain a smaller V_m. It, however, does not hold for all situations where V_m does not contract.

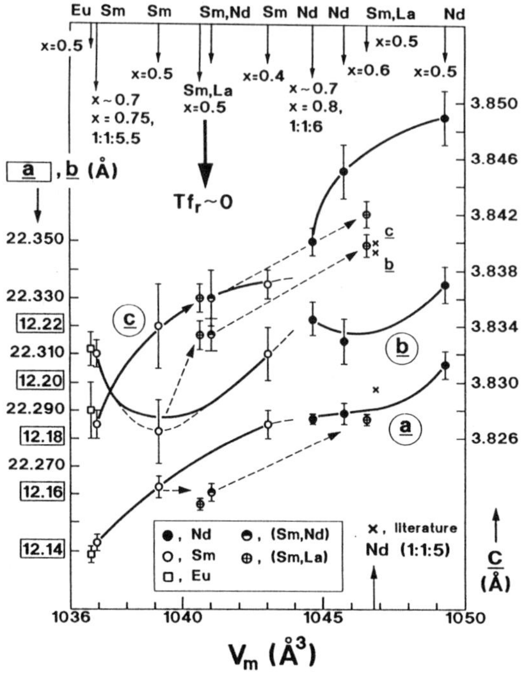

Fig. 9
V_m vs. a, b, c from Ndss through Smss to Euss. Tf_r~0 accomplished by Ln^{3+} substitutions for Sm^{3+} (dashed lines, arrows) at x = 0.5. Unit cells also confirmed from compositions outside of single phase ss region (e.g., 1:1:5.5, 1:1:6, etc.).

Figure 9 revisits the ss. V_m collapses primarily by cooperative contractions of a,b,c. Tf_r decreases accordingly, but subtle details appear. Near terminal Ndss and Smss (x > 0.5), c contracts more rapidly and structural distortions are relaxed by adjustments in a,b. Such behavior seems to be facilitated by an ever increasing concentration of vacancies which must be implicated in ultimate phase instability. Ndss can never attain V_m and A_D necessary for the Tf_r~0 state, being always dominated by +Tf(| |c). Smaller Sm^{3+} and larger Ba^{2+} cooperatively relax distortions and ss stability resumes (x ≥ 0.3) at a V_m close to terminal Ndss. Tf_r is now negative (Fig. 5), being dominated by −Tf(a ∪ b). Tf_r~0 and pertinent A_D are bypassed while the required V_m belongs to a Smss variant of −Tf_r. Similarly, terminal Smss yields to Euss stability where variants are offset in −Tf_r at nearly equivalent V_m. Clearly, ss variants can have identical V_m yet differ in A_D and, hence, Tf_r. If they have identical V_m and A_D, Tf_r may, in fact, differ depending on how the net A_D is accomplished by anisotropic polarizability showing individual variations in a,b,c.

The Smss series initially shows nearly constant −Tf_r, implying that −Tf (a ∪ b) and +Tf(| |c) continuously offset or hardly vary. Within error, variation of b in Fig. 9 is open to interpretation. If b enters a broad minimum, 0.5 > x > 0.4,

$-Tf(||\underline{b})$ could remain nearly constant and, therefore, contribute little added negativity to $-Tf_r$. Also, $\underline{c}$ contracts more slowly suggesting only a small decrease in $+Tf(||\underline{c})$. Beyond x = 0.5, obvious expansion of $\underline{b}$ should impart less negativity, but by now, $\underline{c}$ is into rapid contraction. Thus, $-Tf_r$ near terminal Smss seems to be controlled by polarizability associated with a rapidly decreasing $+Tf(||\underline{c})$ and a decreasing $Tf(||\underline{a})$ that overcome any increase in $-Tf(||\underline{b})$. Similar behavior occurs at x > 0.5 of Ndss, but $\underline{a},\underline{b}$ are restrained from further contraction which cannot cause enough negativity to produce Tf_r~0. Likewise, all x = 0.5Sm,Ln variants display identical behavior when tuned to and through Tf_r~0. Larger Ln^{3+} provide the requisite increase in A_D. Expansion of $\underline{b}$ and a slight increase in $\underline{c}$ at constant $\underline{a}$ yield the expected V_m (Fig. 1). In turn, this suggests that Tf_r~0 is attained by polarizability that increases $-Tf(||\underline{b})$. When x ≠ 0.5, tuning to Tf_r~0 involves different A_D (Table I) and $V_m(\underline{c}$, Fig. 5). Figure 9 also shows x = 0.5Sm,La (Table I, Fig. 5) tuned through Tf_r~0 into the V_m and A_D ranges of Ndss. Clearly, its Tf_r is determined by a combination of polarizabilities associated with $\underline{a},\underline{b},\underline{c}$ that differs from pure Ndss at equivalent V_m.

Figures 10, 11, 12, and Tables II,III give examples of data proving all contentions discussed, as well as additional evidence. Figure 10, derived from ceramics near 1:1:5Sm, shows that PCA and measured EA depend on processing. Here, EA is defined by the absolute difference between measured Tf_N^A, Tf_N^B and $\acute{E}_N^A$, $\acute{E}_N^B$ from the same ceramic. I ceramics always lack anisotropy and H and E produce PCA/EA only above a critical calcine temperature (CCT). XRD proves that this CCT (1025°-1050°C) corresponds with the first appearance of ss as a reaction product. Above CCT, the ss population increases and this correlates with enhanced PCA/EA. After 1200°C (8 hr), powders of Sm and Nd formulations are not fully reacted and have modest concentrations of ss. Corresponding E and H ceramics attain PCA of about 6% (V_1) and 2%, respectively. Thus, relatively small EA, especially for É, is observed. Nevertheless, large intrinsic EA is computed (see below). CCT is not a constant because it depends on physical/chemical properties and reaction processes associated with raw materials selected. The ss, for example, develops in the 975°-1025°C range in formulations containing Pb and/or Bi and by 1150°C powders are almost fully reacted. In turn, measured PCA/EA are much larger (see below). The generic nature of Fig. 10 must be appreciated in other ways. While correct, it cannot be inferred that H culminates in less PCA than E without other supporting evidence since property differentials are pertinent to different types of PCA. Also, the magnitude of property differentials is not constant but can be varied slightly by changing forming variables (e.g., pressure, die design, etc.).

Despite prior contentions (14), Fig. 11 proves that Tf_r (and $\acute{E}_r$) of Ndss is tuned by Bi^{3+} substitutions. C-M easily resolves how Bi^{3+} substitutes. Given measured V_m (see below), if Bi^{3+}(a_D = 6.12Å^3) replaces Ba^{2+} (6.40), $\acute{E}_r$ must decrease. If Bi^{3+}replaces Nd^{3+} (5.01), $\acute{E}_r$ must increase. The data prove the latter and in simplest form the ss is $Ba_{6-3x-z}A_zNd_{8+2x-y}Bi_yTi_{18}O_{54}$. Clearly, Tf_r never attains the zero state. Instead, broad minima occur in the y = 1.0-1.5 range. Beyond these, Tf_r and $\acute{E}_r$ increase accompanied by a rapid decrease in $\underline{Q}$. This behavior may portend the appearance of secondary phase(s) which are first detected by XRD for x = 0.5 at y = 3. On the other hand, V_m continues to vary, albeit in curious fashion, Table II. If correct, an apparent solubility limit of 2 < y < 3 is established. Yet, we note several problems when y > 1.0. Tf_r and V_m show

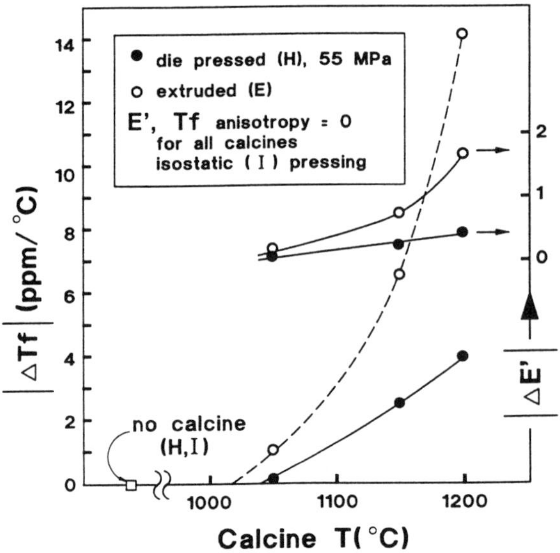

Fig. 10
Influence of forming and calcine temperature on measured (TE field) electrical anisotropy using one blend of raw materials formulated near 1:1:5Sm. Each property differential is the absolute difference of measured values from type A and B samples (Fig. 6) taken from the same ceramic. Anisotropy is observed above a critical calcine temperature.

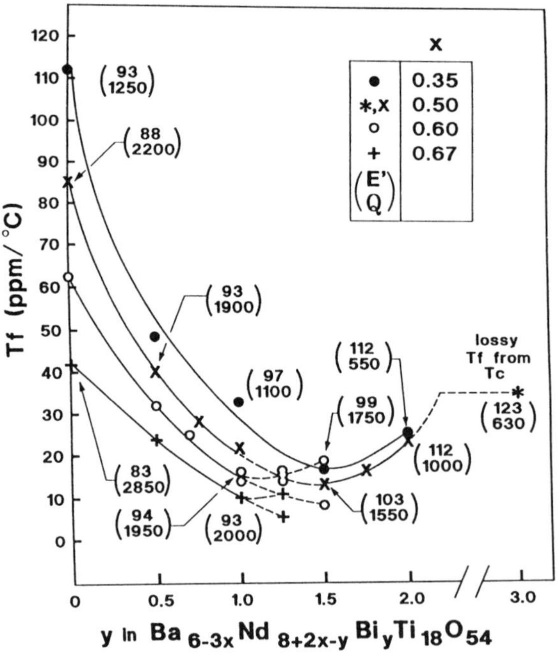

Fig. 11
Tf_r [and example É,Q (~3GHz)] for Bi-substituted Ndss. Limits of Bi^{3+} solubility traced to at least y = 1.0 before encountering difficulties (see text).

significant variability depending on firing temperature. Lower temperatures (e.g. 1300° vs. 1325°-1350°C) favor more consistent correlations with trends in Fig. 11. Vaporization of Bi-oxide may be a complicating factor. Furthermore, pure Ndss melts initially near 1370°C (19). Bi-substituted variants must melt at lower quaternary temperatures. Thus, both processes may be influencing the data. At best, the solubility of Bi can be followed to $1.0 < y \leq 1.5$. Equilibrium secondary phase(s) also will decrease the temperature of quaternary melting. The x = 0.5 composition where Bi^{3+} replaces all Nd^{3+} was examined. The ss is absent and the phase assemblage is dominated by ferroelectric Ba, Bi-titanate. The presence of this lossy, high É phase after cooling would explain the variation of properties beyond the minima. Regardless, the Nd,Bi-system yields ceramics with the best combination of Tf,Q at the highest É reported to date.

TABLE II
Properties of Selected $Ba_{6-3x-z}A_zNd_{8+2x-y}Bi_yTi_{18}O_{54}$ Formulations

x	y	z	a(Å)	b	c	V_m(Å³)	Tf_r(ppm/°C)	$É_r$	Q
0.5	0	0	12.207(5)	22.335(7)	3.849(2)	1049.3(6)	85	88	2210
(a)	0	0.9Sr	12.197(4)	22.319(8)	3.841(2)	1045.6(7)	63	86.7	2380
	0	1.0Pb	12.196(2)	22.326(4)	3.841(1)	1045.8(2)	21	91	1860
	0.5	0	12.198(3)	22.317(5)	3.851(2)	1048.3(4)	40	93.4	1900
	1.0	0	12.209(3)	22.324(6)	3.858(2)	1051.4(4)	22	97	1830
	1.0	0.5Pb	12.197(3)	22.333(5)	3.851(1)	1049.0(2)	12	99	1750
	1.0	1.0Pb	12.205(2)	22.340(3)	3.846(1)	1048.7(2)	-4	101	1290
	1.5	0	12.209(4)	22.330(6)	3.859(2)	1052.0(4)	13	103	1560
	2.0	0	12.206(3)	22.344(5)	3.858(1)	1052.2(3)	25	112	1010
	2.0	1.0Pb	12.211(2)	22.350(5)	3.851(1)	1050.9(3)	-5	115	860
NSP	3.0	0	12.209(4)	22.352(7)	3.858(2)	1052.9(5)	~37	123	630
0.6	0	0	12.189(4)	22.315(8)	3.845(2)	1045.7(6)	62	87	2210
	0.5	0	12.195(2)	22.316(4)	3.851(1)	1048.0(4)	32	89.5	2170
	1.0	0	12.206(4)	22.321(7)	3.852(2)	1049.5(5)	15	94	1980
(a)	1.25	0	12.200(1)	22.323(3)	3.854(1)	1049.5(2)	17	98	1700
(b)	1.25	0	12.196(2)	22.323(3)	3.848(1)	1047.7(2)	14	96	1920
	1.50	0	12.206(3)	22.334(5)	3.854(2)	1050.7(4)	19	98.7	1740
(b)	1.50	0	12.203(2)	22.319(4)	3.849(1)	1048.2(2)	8	98.8	1740

All electrical data near 3GHz. Except for (a,b), XRD data are from one laboratory. (a,b) represent XRD from a second laboratory. Samples (a,b) fired at different temperature (note electricals). NSP=clearly not of single phase (see text).

Table II summarizes data for examples of x,y,z variants. For x = 0.5,y, Tf_r rapidly decreases initially because V_m clearly contracts primarily by a decrease in b. By x = 0.5, y = 1.0, a and c clearly have expanded to produce a net V_m increase. From prior observations, this should cause Tf_r to increase, yet it continues to decrease. By x = 0.5, y = 1.5, a and c are constant and V_m, within error, has hardly changed. This correlates with the Tf_r minimum, Fig. 11. The x = 0.6, y series also shows similar behavior as V_m increases slightly to a maximum corresponding with the Tf_r minimum. While XRD data may be influenced by some loss of compositional and phase integrity (see above), we believe that general trends in a,b,c, and V_m to x,y = 1.5 probably hold. If correct, it is difficult to deduce how polarizability varies to accomplish a net decrease in Tf_r, and, therefore, it is tempting to concede that the Bi^{3+} case is anomalous. However, some interesting features are evident. When V_m expands, it is accompanied by checks and balances in a,b,c. For example, from x = 0.5,y = 0.5 to x = 0.5,y = 1.0, the rise in V_m is checked by b which, within error, may remain constant. Progressing to x = 0.5,y = 1.5, V_m is constrained certainly by constant a and, within error, possibly by b. Initial expansion of x = 0.6,y = 0 through x = 0.6,y = 1.0 is restrained by b. Further increase in V_m within the Tf_r minimum is checked at least by constant a. Clearly, Bi^{3+} substitutions continuously distort the structure in complex ways to accommodate small concentrations. We speculate that constrained a and/or b reflect the negativity necessary to reduce Tf_r despite small increases in c and V_m.

Table II shows examples of how smaller Sr^{2+} and Pb^{2+} change electrical and unit cell properties of a parent ss when substituted for Ba^{2+}. Sr^{2+} (4.24Å^3) reduces net A_D and decreases V_m via cooperative a,b,c contractions such that É diminishes in accord with C-M. Highly polarizable Pb^{2+} (6.58) increases net A_D slightly and always decreases V_m so that $\acute{E}_r$ increases in accord with C-M. Without Bi^{3+}, Pb^{2+} decreases a,b,c similar to Sr^{2+} at nearly equivalent concentration. With Bi^{3+}, Pb^{2+} contracts c, but inherent distortion due to Bi^{3+} is adjusted by further changes in a,b. Both Sr^{2+} and Pb^{2+} reduce Tf_r, but the latter is more effective, supporting empirical findings (15). In the presence of Bi^{3+}, Pb^{2+} is less effective because contractions in c are countered by expansions in a,b that limit negativity.

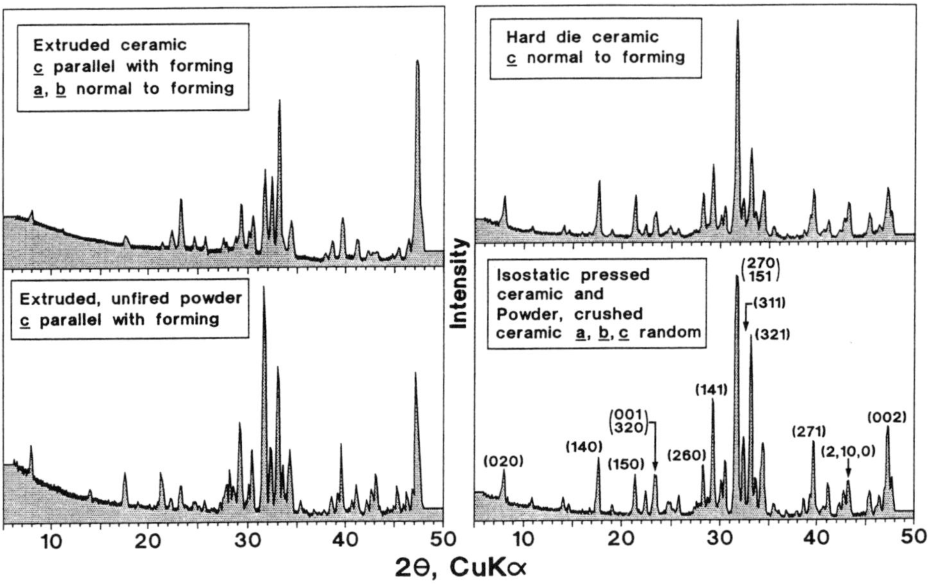

Fig. 12 Variation of XRD intensities due to random and preferred crystallite alignments imparted by forming and enhanced during densification. Example (hkℓ) are given ignoring doubled c. The composition is x = 0.6, y = 1.0.

Figure 12 illustrates XRD data for one example, x = 0.6,y = 1.0. Equilibrated powder (1250°C) and I ceramics yield XRD intensities of the random state and these are in excellent accord with powder data in (9,16). Critical (00ℓ) intensities used in Lotgering calculations reveal that the E ceramic has PCA ~16%. The H ceramic displays smaller PCA~5%. In fact, PCA initiated by E can now be discerned in the

TABLE III
Electrical Properties of Selected $Ba_{6-3x}Ln_{8+2x-y}Bi_yTi_{18}O_{54}$ Formulations

x	y	Forming	Sample Type	É 3GHz	É 1MHz	T*f* 3GHz	Tc(T*f*) 1MHz	V_1
0.5Sm	0	I	A,B	81.3	81	-14	29 (-19)	0
		H*	A	80.6	78.9	-7.5	52 (-30)	0.02
			B	79.8	79.6	-16	18 (-14)	
		E	A	80.5	83.8	-26	-31 (11)	0.06
			B	82.6	80.4	-7	63 (-36)	
0.5Nd	0.5	I	A,B	93.4	93	40	-95 (43)	0
		H	A	94.6	91.7	47	-41 (16)	0.05
			B	93	94	35	-88 (40)	
0.6Nd	0	I	A,B	87	87	62	-122 (56)	0
0.6Nd	1.0	I	A,B	94.3	--	15	--	0
		H	A	95.8	91.6	26	30 (-20)	0.05
			B	94	96	4	-52 (21)	
		E	A	89	101	-22	-136 (63)	0.16
			B	95	90	25	50 (-30)	

For Tc(T*f*), T*f* computed via eq. (2); both as ppm/°C. *, H samples of lower density (5.69) than I, E (5.73g/cc). V_1=estimated fraction of oriented crystallites (see text).

green compact prepared from powder reacted at 1200°C. In contrast to Sm and Nd formulations, this powder contains a much larger population of ss. E and H orient more of the available ss, and other processes during densification enhance PCA/V_1. Thus, measured EA from these ceramics is much larger than from Sm, Nd ceramics. Calculated anisotropic properties for each ss, however, are similar.

Table III summarizes EA for several ceramics. Tf_r and $\acute{E}_r$, independent of field and sample type (Figs. 6,8), are obtained only from I. Properties from all other samples differ for reasons discussed. Bi-substituted variants show much larger anisotropy than x = 0.5Sm and 1:1:5Sm (Fig. 10) because of larger PCA/V_1. Given eq. 2 and necessarily crude models defined by eqs. 4,5, the data clearly demonstrate sign and magnitude of <u>intrinsic</u> EA within constraint of errors as discussed. In evaluating eq. 5 for H and E, care must be taken to use $\acute{E}_r$ from an I ceramic of similar density. If unavailable or unobtainable, $\acute{E}_N^B$ can be substituted if V_1 is small because PCA is almost random. $Tf(\underline{a} \cup \underline{b})$ and $\acute{E}(\underline{a} \cup \underline{b})$ are easily approximated from eqs. 4,5 expressed as (4)$Tf_E^A \sim V_1Tf(\underline{a} \cup \underline{b})+(1-V_1)Tf_r$ and (5)$\log \acute{E}_E^A \sim V_1\log\acute{E}(\underline{a} \cup \underline{b})+(1-V_1)\log \acute{E}_r$. All data show that $Tf(\underline{a} \cup \underline{b})$ is more <u>negative</u> than at least 200-250 ppm/°C. For all variants within about ±60 ppm/°C of Tf_r~0, $\acute{E}(\underline{a} \cup \underline{b})$ is <70. $Tf(\| \underline{c})$,$\acute{E}(\| \underline{c})$ are obtained from type A samples of E using Ci, Figs. 6,8. For example, eq. 4 takes the form of Tf_E^A (from Tc_E^A) $\sim V_1Tf(\| \underline{c})+(1-V_1)Tf_r$. All data show that $Tf(\| \underline{c})$ is <u>positive</u> at a <u>minimum</u> of several hundred ppm/°C. $\acute{E}(\| \underline{c})$ from $\acute{E}_E^A$ using Ci and eq. 5 always is >140. Similar calculations using Tf_H^A,$\acute{E}_H^A$ which contain information from $Tf(\underline{a} \cup \underline{c})$, $\acute{E}(\underline{a} \cup \underline{c})$ support these findings. Note, for example, that Tf_H^A is always $>Tf_r$ because of the influence of $Tf(||\underline{c})$. At equivalent density, $É_H^A$ also must be $>É_r$ because of the large contribution from $É(||\underline{c})$.

Type A specimens from H elucidate $Tf(||\underline{b})$,$É(||\underline{b})$ when interrogated by Ci (Fig. 8). $Tf(||\underline{b})$ is always highly negative and well over 400-500 ppm/°C. For $Tf_r \sim 0\pm60$, $É(||\underline{b})$ is ≤60. Clearly, $\underline{b}$ dominates both $Tf(\underline{a} \cup \underline{b})$ and $É(\underline{a} \cup \underline{b})$. Unfortunately, properties associated with $\underline{a}$ cannot be isolated experimentally and simple averaging cannot be used to extract properties from property unions. However, $\underline{a}$ properties must fall in the sequences, $-Tf(||\underline{b})<Tf(||\underline{a})<+Tf(||\underline{c})$ and $É(||\underline{b})<É(||\underline{a})<<É(||\underline{c})$. We suspect that $Tf(||\underline{a})$ is small and possibly negative.

An understanding of PCA/EA allows for some flexibility in tuning properties via processing. If the H process is selected, $Tf_H^A = Tf_r$ at $V_l = 0$. When V_l is finite, $Tf_H^A > Tf_r$ and can never be zero or negative unless Tf_r is first made negative by chemical adjustment. However, the E process can yield $Tf_E^A << +Tf_r$ because of $-Tf(\underline{a} \cup \underline{b})$. As V_l decreases in eq. 4, Tf_E^A approaches Tf_r. At some V_l, depending on Tf_r and $-Tf(\underline{a} \cup \underline{b})$ magnitudes, Tf_E^A must cross zero. In Table III, the Nd,Bi ss has done so. To make Tf_E^A less negative, V_l (i.e., PCA) must be reduced by decreasing the population of ss in the powder, implicit in Fig. 10, and/or by retarding mechanisms that enhance V_l during firing. When $+Tf_r$ is large and, therefore, requires a large increase in V_l to obtain compensation from $-Tf(\underline{a} \cup \underline{b})$, tuning capability is constrained because the E process seems limited to $V_l<20\%$ even when powder is nearly 100% ss.

III. SUMMARY AND CONCLUSIONS

Properties measured from ceramics of the ss do not reflect bulk chemistry uniquely since processing details cause variations. É for any proposed chemistry should be tested for validity using C-M. Unfortunately, definitive structural data including variable chemistry are not available to help develop a bonding model that explains electrical properties. Some concepts are amplified below and valuable insight can be gained from property trends associated with titanate and zirconate perovskites of Ca, Sr, Ba, and Pb.

$É_r$ consists of polarizability components that produce large $É(||\underline{c})$ and smaller $É(\underline{a} \cup \underline{b})$. Likewise, Tf_r is the resultant of large $+Tf(||\underline{c})$ and $-Tf(\underline{a} \cup \underline{b})$ and, therefore, the components of $É_r$ decrease and increase, respectively, with increasing temperature. Such observations implicate multiple bonding mechanisms. The sign of Tf_r, however, may or may not reflect which dominates since the directional dependence of Tf is complicated further by the tensor nature of polarizability. For example, $+Tf(||\underline{c})$ may be affected by polarizability changes along $\underline{c}$, as well as changes in $\underline{ab}$. In addition, data show that mechanisms controlling V_m (i.e., thermal expansion and substitutions) produce different changes in É and, therefore, polarizability.

Bonding of Ti^{4+} to tilted, corner-shared oxygens along $\underline{c}$ probably dominates $É(||\underline{c})$, but it is unclear whether it involves directionally critical ion displacements that signal spontaneous polarization (22) or its onset. Substitutions of larger cations in the edge-shared polyhedra increase V_m but inherently add polarizability. Thus, it is difficult to discern if the Ti^{4+}-oxygen polarizability also intensifies.

However, replacement of Ti^{4+} by larger Zr^{4+},Sn^{4+} also expands V_m and adds polarizability. Yet, $É_r$ (and Tf_r) decreases (19) suggesting, at least, that details of the Ti^{4+}-oxygen bond are critical.

Net polarizability along c governs properties from Nd through La since Tf_r, dominated by $+Tf(||c)$, becomes increasingly positive. From Nd through Eu, $É(||c)$ remains large but $+Tf(||c)$ decreases. At Ba-rich Smss, Tf_r switches from positive to negative character indicating dominance of $-Tf(a \cup b)$. Neither Ndss or Smss has the correct balance of polarizabilities to produce $Tf_r \sim 0$. It is obtained by substituting larger Ln^{3+} in Smss. These increase V_m slightly, but, more importantly, cause subtle variations in a,b,c. At x = 0.5Sm, for example, $-Tf(a \cup b)$ acquires less negativity because of minor bonding adjustments manifested by a large increase in b while a,c remain constrained. This and similar patterns are repeated frequently as illustrated by Ln-rich ss members and by substitutions of Bi^{3+} or of Bi^{3+} plus Pb^{2+}.

$É(a \cup b)$ must be dominated by polarizability involving the edge-shared polyhedra. Uniform collapse of V_m and/or independent contractions of and constraints on a,b suggest that oxygen and large cations are deformed. The process is facilitated by larger concentrations of vacancies; by smaller, "harder sphere" Sr^{2+}; and by smaller Ln^{3+}. All decrease V_m and inherently diminish calculated polarizability. $É(a \cup b)$ and $É(||c)$ both decrease, but $-Tf(a \cup b)$ exerts greater influence. Tf_r decreases most rapidly when highly deformable Bi^{3+} and/or Pb^{2+} are present. Being smaller than Ba^{2+},Pb^{2+} collapses V_m and, therefore, adds its deformation to the overall. Bi^{3+} is slightly larger than Nd^{3+}, but also adds its deformation and/or influences that of other ions in accordance with observed variations in a,b,c. Both cations add sufficient polarizability to increase $É_r$, yet greater negativity is imparted to $-Tf(a \cup b)$. Alternatively, less positive Tf_r could be the result of a rapidly decreasing $+Tf(||c)$. Alignment of the lone pair parallel with c also may have special importance.

Thermal expansion produces seemingly contradictory properties, but it is difficult to deduce what combinations of complex thermal motion are responsible without good anisotropic parameters. Greater rigidity of polyhedra along a,b is consistent with the concept of counter-vibrating, deformed ions that add polarizability (18). Large expansion of c probably reflects stretching of Ti^{4+} bonds to corner-shared oxygens and some relaxation of the tilt. This may account for the decrease in attendant polarizability. However, the tilted oxygens may prove decisive in coupling polarizabilities of polyhedra and octahedra. In fact, they are most susceptible to thermal motion and to changes in V_m due to substitutions. If correct, the general interpretations of $É_r$ and Tf_r as advanced may require further modification. Here, evidence from ss members containing Ln^{3+} larger than Nd^{3+} and from accurate axial ratios would be invaluable.

IV. REFERENCES

1. R. L. Bolton, Dissertation, Univ. Illinois, 1968.
2. D. Kolar, Z. Stadler, S. Gaberscek, and D. Suvorov, *Ber. Dtsch. Keram. Ges.*, **55**, 346 (1978).
3. D. Kolar, S. Gaberscek, B. Volavsek, H. S. Parker, and R. S. Roth, *J. Solid State Chem.*, **38**, 158 (1981).
4. S. Nishigaki, H. Kato, S. Yano, and R. Kamimura, *Am. Ceram. Soc. Bull.*, **66**, 1405 (1987).

5. J. M. Wu and M. C. Chang, *J. Am. Ceram. Soc.*, **73**, 1599 (1990).
6. J. W. Sun, C. C. Wei, and L. Wu, *J. Mater. Sci.*, **27**, 5818 (1992).
7. R. G. Matveeva, M. B. Varfolomeev, and L. S. Il'yushchenko, *Russ. J. Inorg. Chem. (Engl. Transl.)*, **29**, 17 (1984).
8. A. M. Gens, M. B. Varfolomeev, V. S. Kostomarov, and S. S. Korovin, *Russ. J. Inorg. Chem. (Engl. Transl.)*, **26**, 896 (1981).
9. J. Takahashi, T. Ikegami, and K. Kageyama, *J. Am. Ceram. Soc.*, **74**, 1873 (1991).
10. J. P. Guha, *J. Am. Ceram. Soc.*, **74**, 878 (1991).
11. T. Jaakola, A. Uusimäki, R. Rautioaho, and S. Leppävuori, *J. Am. Ceram. Soc.*, **69**, C-234 (1986).
12. H. Ohsato, S. Nishigaki, and T. Okuda, *Jpn. J. Appl. Phys.*, **31**, 3136 (1992).
13. F. Beech, K. Davis, A. Santoro, R. S. Roth, J. L. Soubeyroux, and M. Zucchi, Ann. Mtng. Am. Ceram. Soc., Abst. 251-B-87, 1987. See also 14th Int'l. Cong. of Cryt., Abst. 07.9-9, 1987.
14. D. Kolar, S. Gaberscek, Z. Stadler, and D. Suvorov, *Ferroelectrics*, **27**, 269 (1980).
15. K. Wakino, K. Minai, and H. Tamura, *J. Am. Ceram. Soc.*, **67**, 278 (1984).
16. Standard X-ray Diffraction Powder Patterns, NBS Monograph 25, Sec. 19, p. 19, 1982.
17. R. D. Shannon, *J. Appl. Phys.*, **73**, 348 (1993).
18. P. J. Harrop, *J. Mater. Sci.*, **4**, 370 (1969).
19. T. Negas, Internal Trans-Tech Report 124292, Trans-Tech, Inc., 1992 (edited version to be published).
20. T. Negas, G. Yeager, S. Bell, and R. Amren, Chemistry of Electronic Ceramic Materials, NIST Spec. Pub. No. 804, p. 21 (1991).
21. F. K. Lotgering, *J. Inorg. Nucl. Chem.*, **9**, 113 (1959).
22. J. P. Mercurio, M. Manier, and B. Frit, *Ferroelectrics*, **127**, 35 (1992).
23. Y. M. Poplavko, M. E. Ilchenko, and V. I. Butko, 23rd Eur. Micro. Conf. Proc., 600 (1993).

V. APPENDIX

CCT = critical calcine temperature; first appearance of ss.

$\acute{E}_r$ (random ss alignment) = $\acute{E}_I^M$ = $\acute{E}(\underline{a} \cup \underline{b} \cup \underline{c})$.

EA = electrical anisotropy measured from ceramic or intrinsic to the ss as calculated.

Forming: I = isostatic, H = hard die, E = extrusion.

M = sample type, N = forming method for properties Tf_N^M, Tc_N^M, $\acute{E}_N^M$, etc.

PCA = preferred crystallite alignment.

$Tf(\|\underline{c})$, $\acute{E}(\|\underline{b})$, etc. = properties parallel with axis shown.

$Tf(\underline{a} \cup \underline{b})$, $\acute{E}(\underline{a} \cup \underline{c})$, etc. = property unions pertinent to crystallographic axes.

Tf_r (random ss alignment) = Tf_I^M = $Tf(\underline{a} \cup \underline{b} \cup \underline{c})$.

V_1 = volume fraction of aligned crystallites.

CHEMISTRY AND MICROWAVE PROPERTIES OF Bi_2O_3 DOPED $Ba_{4.5}Nd_9Ti_{18}O_{54}$ BASED CERAMICS

D. Suvorov, M. Valant, D. Kolar
"Jožef Stefan" Institute, University of Ljubljana, Jamova 39, 61000 Ljubljana, Slovenia

ABSTRACT

Ceramics based on the $BaO-Nd_2O_3-TiO_2$ system are widely used in the electronics industry due to their high permittivity and low dielectric losses. An especially interesting part of the phase diagram is the solid solubility region with the general formula $Ba_{6-x}Nd_{8+2/3x}Ti_{18}O_{54}$ (O < X < 2.25). For microwave applications such materials are doped with Pb or Bi oxides (or titanates) to compensate the relatively high temperature coefficient of resonant frequency τ_f. In the present paper, a study of Bi_2O_3-doped 1:1:4-BT_4 ceramics is reported with regard to the formation of solid solution and liquid phase during sintering, and the microstructural development and its influence on the microwave properties. It was found that on addition of up to 2.5 mol % Bi_2O_3, Bi^{3+} is incorporated on Nd^{3+} sites in the $Ba_{4.5}Nd_9Ti_{18}O_{54}$ phase. When the solid solubility limit is exceeded, Bi_2O_3 concentrates at the grain boundaries as Bi-rich phase, forming a liquid phase and causing a considerable decrease of the Q-value and an increase of τ_f.

INTRODUCTION

The investigation of ceramics based on $BaO-Ln_2O_3-TiO_2$ systems was stimulated by the observation that ceramics based on compositions in the TiO_2-rich corner of these systems exhibit a highly temperature stable permittivity and low losses, as required for the manufacture of type 1 ceramic capacitors. According to Negas et al.(1), the first extensive study of the phase relationships in such systems was performed by Bolton (2). By varying the compositions when using Nd_2O_3 he prepared a two phase ceramic (X and Y phase) with low losses (at

1 MHz), a small temperature coefficient of capacitance (T_c) and high E′ (over 60). Later, over a decade, it was shown in detail that dielectric ceramics based on $BaO-Ln_2O_3-TiO_2$ systems near the composition $BaO:Ln_2O_3:4TiO_2$ (show interesting combinations of high dielectric constant (110-80), low losses (tg$\delta \leq 1\text{-}2.10^{-4}$) and a low T_c, which classify them as potential materials for electronic applications (3-5). Kolar et al.(6) characterized the electrical properties (at 1 MHz) of the system containing Nd_2O_3, showing that compositions around 1:1:5 - 1:1:4 exhibit greatly improved properties for NPO-type ceramics. To increase the T_c value of such ceramics (from approximately - 120 ppm/K to the range from 0 to ± 30 ppm/K), bismuth titanate was used as additive (7).

It was natural that the good dielectric properties of 1:1:5 - 1:1:4 type compositions at 1 MHz attracted the attention of engineers and scientists in search of materials for use in devices operating in the GHz range. In adapting 1:1:4 type ceramics for microwave applications, the main tasks are to retain the high permittivity and low losses while tuning τ_f to zero. Two approaches are used: improvement of τ_f by additives, or/and by a proper combination of rare earth elements in the 1:1:4 solid solutions.

Popular among the additives are PbO, Bi_2O_3 (Bi-titanates) and TiO_2. Wakino et al.(8) described 1:1:5 Nd-containing ceramics in which, by substituting some Ba by Pb, they achieved a high E′ (85-88), a modest Q (≈ 2000 at 3 GHz) and a near zero τ_f. Nishigaki et al. reported (9) that the Sm 1:1:5 composition ceramics exhibit a high E′ (80) and high Q (3700 at 3 GHz), but negative τ_f. By substitution of a small amount of Ba by Sr, they were able to tune τ_f through zero. The same tuning of τ_f in the 114 Sm system was achieved by Negas (1) simply by adjusting the TiO_2 content along the join TiO_2 - 1:1:4.

The beneficial influence of Bi_2O_3 addition in improving τ_f and promoting densification of 1:1:4 type ceramics has been reported by several authors (1, 10-12). However, explanations of the role of Bi_2O_3 leading to the improvement of dielectric properties are incomplete and in disagreement.

In earlier studies (7), the beneficial influence of Bi_2O_3 addition was ascribed to its ferroelectric nature, showing a high dielectric constant (over 100) and a large positive T_c, ≈ 650 ppm/K. 1:1:5 - 1:1:4 type Nd containing ceramics to which 10 wt % bismuth titanate was added exhibited an increase of T_c from -120 ppm/K to -23 ppm/K and slight increase of E′ as compared with the same ceramics without bismuth titanate addition. Present investigation has shown that the reason for the change in properties is replacement of neodimium by bismuth ions in the 1:1:4 lattice. The Bi-modified ceramics showed a two phase structure with a Bi and Ti rich phase between the 1:1:4 grains. Durand and Boilot (11) reported that the 1:1:4 Nd ceramic containing 10 wt % $2Bi_2O_3.3TiO_2$, exhibited a high E′ (≈ 88), a modest Q (≈1500) or Qxf ≈ 5500 (f_r = 4 GHz) and τ_f = 8 ppm/K. Mudrolubova et al.(10) described the limited solid solubility region in the $BaNd_2Ti_4O_{12}$ - $Bi_2O_3.3TiO_2$ system up to ≈ 10 mol % $Bi_2O_3.3TiO_2$, accompanied by an increase of T_c from -100 ppm/K to -20 ppm/K of such a solid solution. Wersing (12) investigated the influence of Bi content on E′, τ_f and Q of 1:1:5 Nd ceramics in more detail and confirmed that Bi addition results in a remarkable decrease in τ_f, and at the same time, even increases E′. The Q factor slowly decreased with Bi content, as a consequence of the second phase. By analogy with $BaBiO_3$, he proposed that bismuth could substitute Ti in 1:1:4 compound. The optimum combination of results was E′ ≈ 95, Q = 2500 (at 2,5 GHz) and τ_f ≈ 10 ppm/K.

Summarizing the reported results, one can conclude that they are consistent in improvement of dielectric properties but much less so with regard to the role of Bi_2O_3 in leading to such improvements. Therefore in this work we investigated in detail the incorporation of Bi^{3+} into the crystal structure of the 1:1:4 Nd (hereafter BNT_4) solid solution, and correlated the results with the dielectric properties.

EXPERIMENTAL

Polycrystalline samples were synthesized by solid state reaction of the corresponding oxides. Reagent grade Bi_2O_3, $Nd_2O_3.2TiO_2$, $BaTiO_3$ and TiO_2 were wet mixed in the proportions described in Table 1 and then calcined at 1150° for 6 hours. After milling the calcined materials to a particle size around 1 μm, the powders were uniaxially pressed into discs under a pressure of 100 MPa and sintered at 1320° for 15 minutes in air.

The microstructures of samples prepared in this way were characterized by scanning electron microscopy (JEOL JXA 840 A). Quantitative analysis was performed by wave dispersion spectroscopy

(TRACOR-NORTHERN) and the phases present identified by X-ray diffraction (ENRAF NONIUS FR 590).

Microwave dielectric properties were measured by a Network Analyzer (HP 8719 C), applying the reflection closed resonant cavity method (13).

RESULTS AND DISCUSSION

All the literature data quoted, as well as the results of our experiments on the influence of Bi_2O_3 addition to the 1:1:4 Nd- containing solid solution can be summarized as shown in Fig. 1.

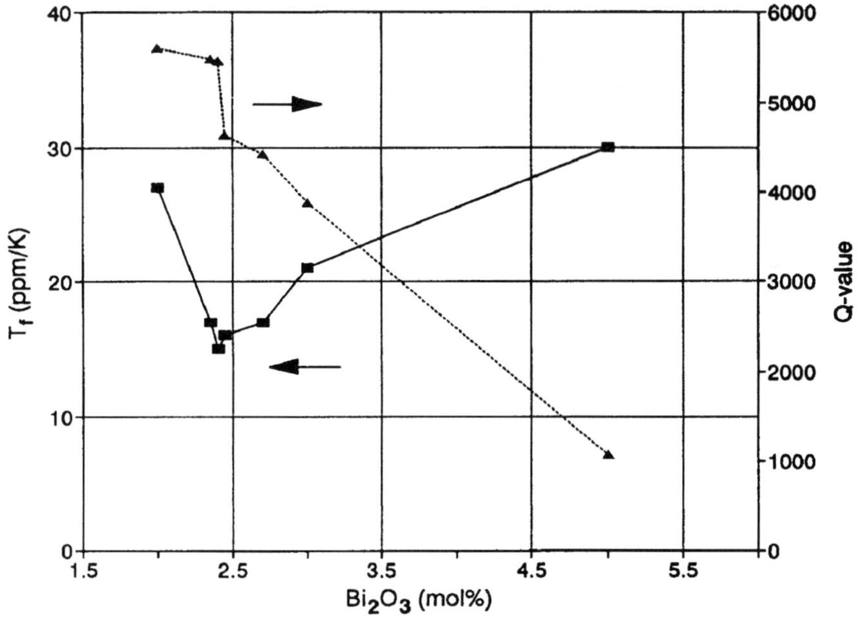

Figure 1. Microwave dielectric properties of ceramics based on BNT4-BT4 system with various contents of Bi_2O_3

One can notice that up to a certain concentration of Bi_2O_3, τ_f gradually decreases and the Q-value remains almost constant. Further addition of Bi_2O_3 sharply reverses the τ_f = f (Bi_2O_3 conc.) trend, indicating a strong increase in τ_f of such ceramics, accompanied by a noticeable worsening of the Q-value. Such observations lead us to the conclusion that the changes in dielectric properties must be strongly correlated with the limited incorporation (i.e. solid solubility) of Bi^{3+} into the BNT_4 matrix.

In our study the basic composition was shifted slightly from the $Ba_{6-x}Nd_{8+2/3x}Ti_{18}O_{54}$ solid solubility region towards BT_4 (92 mol% BNT_4 - 8 mol% BT_4), because preliminary experiments showed that densification of the ceramic with such composition is improved with regard to pure BNT_4.

Three different possible mechanisms for the incorporation of Bi^{3+} in the BNT_4 matrix were studied in detail. Bi^{3+} can theoretically enter BNT_4 at Ba^{2+}, Ti^{4+} (aliovalent substitution) or at Nd^{3+} sites (isovalent substitution). The largest difference in ionic radii is between Bi^{3+} and Ti^{4+} (1.20:0.68 Å) making such substitution less probable. However, Wersing (12) proposed the substitution of Ti^{4+} by Bi^{3+} and Bi^{5+} in 1:1:4 type compisitions according to an analogy with $BaBiO_3$. To verify such a possibility, we prepared ceramics in which Ti^{4+} was partially substituted by Bi^{3+} (Table I).

Table I: Composition of starting powders for the preparation of ceramics in which various oxides were substituted by Bi_2O_3

	BaO (mol%)	Bi_2O_3 (mol%)	Nd_2O_3 (mol%)	TiO_2 (mol%)
Basic composition (a)	16.9	-	15.6	67.5
$Bi^{3+}\rightarrow Nd^{3+}$(a')	16.9	2.0	13.6	67.5
$Bi^{3+}\rightarrow Ti^{4+}$(b)	17.3	2.0	15.9	64.8
$Bi^{3+}\rightarrow Ba^{2+}$(c)	13.1	2.0	15.9	69.0

Microstructural analysis of sintered samples showed that in this case the BT_4 phase completely disappeared and the phase composition was moved into the region between the $BaNd_2Ti_3O_{10}$ and $Ba_{6-x}Nd_{8+2/3x}Ti_{18}O_{54}$ phases (Fig. 2; line a-b, Fig. 3). Such a compositional shift in the direction away from TiO_2 can occur only due to the deficiency of Ti^{4+}, which was not compensated by Bi^{3+} incorporation.

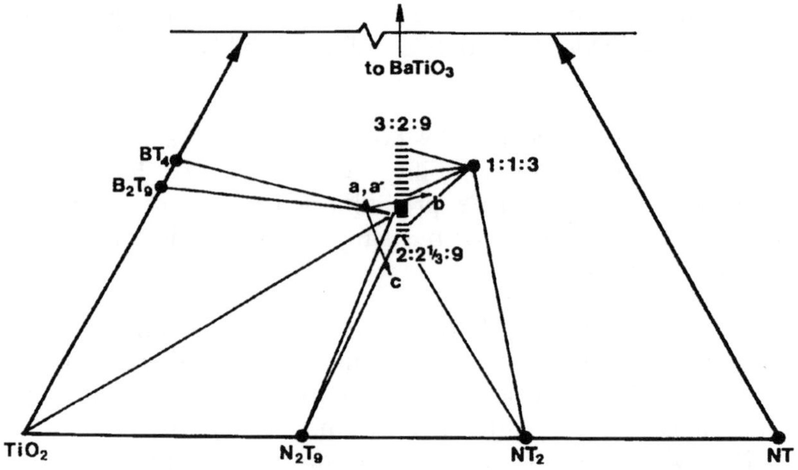

Figure 2. The $BaTiO_3$-TiO_2-Nd_2TiO_5 system according to (1) with indicated compositions investigated in the present work:
(a) basic composition,
(a′) composition after Nd^{3+} was substituted by Bi^{3+},
(b) composition shift after Ti^{4+} was substituted by Bi^{3+},
(c) composition shift after Ba^{2+} was substituted by Bi^{3+}

Microstructural analysis of the sample where Ba^{2+} was substituted by Bi^{3+} also confirms the changes in phase composition (Fig. 4). At least three phases were identified in such samples, indicating a shift from the basic composition into the region between the $Ba_{6-x}Nd_{8+2/3x}Ti_{18}O_{54}$ - $Nd_4Ti_9O_{24}$ - $Nd_2Ti_2O_7$ phases (Fig. 2; line a-c). Therefore we can conclude that Bi^{3+} does not replace Ba^{2+} either.

Microstructural analysis of the BNT_4 - BT samples in which up to 2.4 mol% Nd_2O_3 had been substituted by Bi_2O_3, confirmed the presence of two phases, BNT_4 and BT_4, identically to the starting composition without Bi_2O_3 addition (Fig. 5, Fig. 2; point a, a′). WDS analysis revealed Bi^{3+} in the BNT_4 grains. BT_4 grains were identified as Bi-free. Even after prolonged sintering of such ceramics the phase compositions and morphology of the phases remained unchanged, leading to the conclusion that such behavior is possible only when Bi^{3+} cations substitute on the mixed Ba/Nd/vacancy crystallographic position.

Further addition of Bi_2O_3, over 2.4 mol% caused changes in the morphology of the BT_4 phase regions. At concentrations higher than 5 mol % Bi_2O_3 an additional low melting Bi-rich phase was found by SEM on the grain boundaries of the matrix. The presence of liquid phase during sintering of such ceramics slightly increased the grain growth of the BNT_4 matrix grains (Fig. 6).

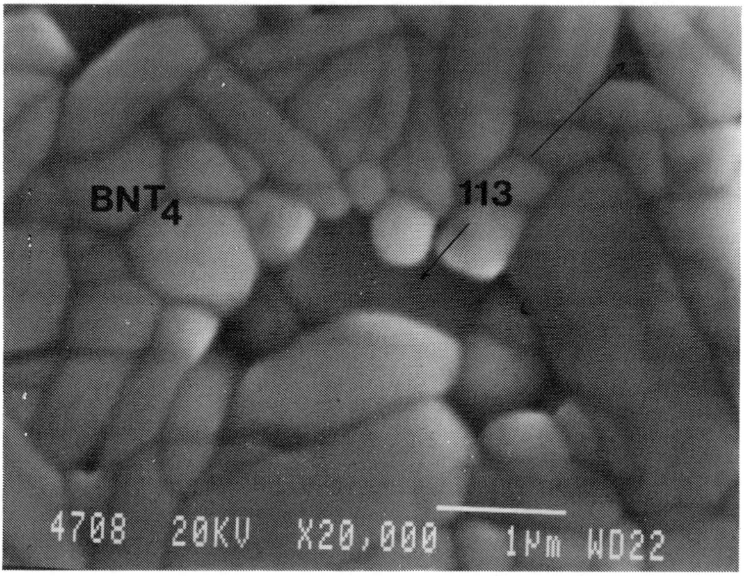

Figure 3. Microstructure of thermally etched BNT_4-BT_4 based ceramics ($Bi^{3+} \rightarrow Ti^{4+}$ substitution), sintered at 1320°C for 15 minutes.

As mentioned above, quantitative WDS analysis of the BNT_4 phase in the samples containing Bi_2O_3 confirmed the deficiency of Nd^{3+} and the presence of Bi^{3+} in the BNT_4 grains when up to 2.5 mol% Nd_2O_3 was substituted by Bi_2O_3. Assuming Bi^{3+} substitution of Nd^{3+}, the solid solution can be written as $Ba_{4.5}(Nd_{1-y}Bi_y)_9Ti_{18}O_{54}$. The maximum content of Bi^{3+} detected in the BNT_4 phase was y = 0.15 and therefore the limiting solid solubility may be described by the composition $Ba_{4.5}(Nd_{0.85}Bi_{0.15})_9Ti_{18}O_{54}$. When calculated on oxide contents, this amount of Bi_2O_3 corresponds to 2.5 mol%. As shown in Fig. 1, this concentration of Bi_2O_3 also corresponds to the change of the slope τ_f = f(Bi_2O_3 conc.).

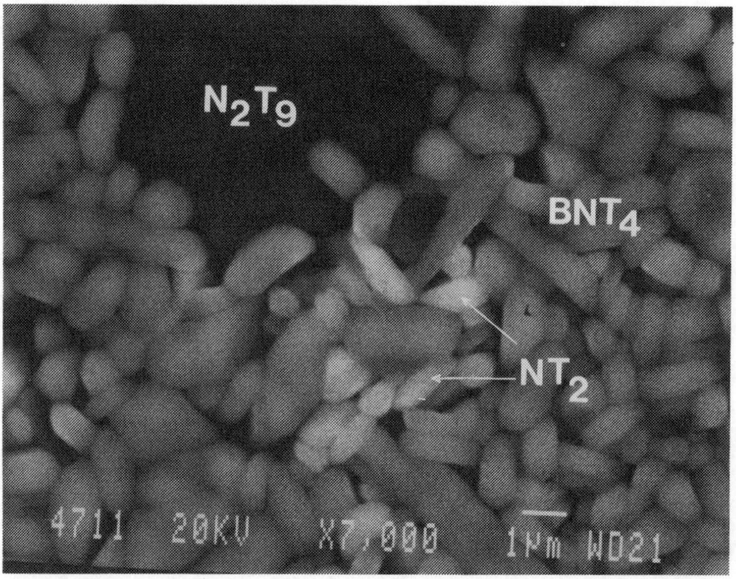

Figure 4. Microstructure of thermally etched BNT4-BT4 based ceramics ($Bi^{3+} \rightarrow Ba^{2+}$ substitution), sintered at 1320°C for 15 minutes.

Figure 5. Microstructure of thermally etched BNT4-BT4 based ceramics ($Bi^{3+} \rightarrow Nd^{3+}$ substitution), sintered at 1320°C for 15 minutes.

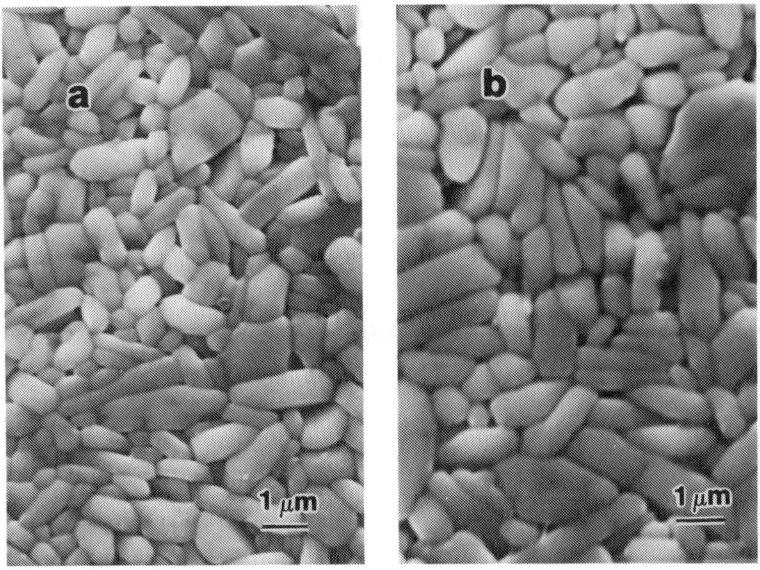

Figure 6. Microstructure of thermally etched BNT4-BT4 based ceramics sintered at 1320°C for 15 minutes
(a) with 2,4 mol% Bi_2O_3
(b) with 5,0 mol% Bi_2O_3

The influence of Bi_2O_3 addition on the microwave properties of $0.92Ba_{4.5}(Nd_{1-y}Bi_y)_9Ti_{18}O_{54}$-$0.08BaTi_4O_9$ based ceramics is given in Table II. One can notice that up to 2.4 mol % Bi_2O_3 addition (i.e. up to the solid solubility limit) τ_f decreases and the permittivity slightly increases, while no significant influence on the Q value can be observed. The best optimization of τ_f (15 ppm/K) was achieved with a sample in which 15 mol % of Nd^{3+} was substituted by Bi^{3+}. Increasing the amount of Bi_2O_3 added (over the solid solubility limit) worsens the τ_f, E′ and Q-values of such ceramics due to the appearance of a Bi-rich phase at the grain boundaries. Further improvements of τ_f to values close to zero must therefore be achieved with other additives (14).

Table II: The influence of Bi_2O_3 addition on the microwave properties of $0.92Ba_{4.5}(Nd_{1-y}Bi_y)_9Ti_{18}O_{54}$ - 0.08 $BaTi_4O_9$ based ceramics

Bi_2O_3 addition (mol %)	y*	f_r (GHz)	E′	τ_f (ppm/K)	Q (at f_r)
2.00	0.120	4.3016	96	27	1302
2.35	0.145	4.2088	98	17	1300
2.40	0.150	4.1877	99	15	1301
		Solid solubility limit			
2.45		4.1801	99	16	1108
2.70		4.2275	97	17	1046
3.00		4.1409	91	21	937
5.00		4.8899	75	30	219

*y in $Ba_{4.5}(Nd_{1-y}Bi_y)_9Ti_{18}O_{54}$

CONCLUSIONS

Bi_2O_3 interacts with $Ba_{4.5}Nd_9Ti_{18}O_{54}$ and forms a solid solution. Bi^{3+} replaces Nd^{3+} ions in the crystal structure of the 1:1:4 Nd type phase and this solid solution may be represented by the formula $Ba_{4.5}(Nd_{1-y}Bi_y)_9Ti_{18}O_{54}$. The solid solubility limit was determined to be at y = 0.15 (2.5 mol % Bi_2O_3).

Further increase in the amount of Bi_2O_3 addition causes the formation of a Bi-rich phase which concentrates at the grain boundaries of the ceramic, leading to a considerable decrease of the Q-value and an increase of τ_f.

The optimal improvement of microwave properties was achieved with addition of 2.4 mol % Bi_2O_3 to 0.92 BNT_4 - 0.08 BT_4 ceramics, resulting in a permittivity of 99, Q x f ≈ 5500 and τ_f = 15 ppm/K.

LITERATURE

1. T. Negas, G. Yeager, S. Bell, R. Amren, Chemistry and Properties of Temperature Compensated Microwave Dielectrics, pp. 27 in Chemistry of Electronic Ceramic Materials (Ed. by P. K. Davis and R. S. Roth), NIST Spec. Publication 804 (1991)
2. R. L. Bolton, Dissertation, University of Illinois 1968

3. E. A. Nenasheva, B. A. Rotenberg, E. J. Gindin, V. G. Prohvatilov, Dielectric Properties and Structure of Rare Earth Titanates and Alkaline Earths at Isovalent Substitutions in Various Sublattices, Neorg. Mat. 16 (6), (1980), 1040-1043
4. L. P. Mudrolubova, Investigation of Physico-Chemical Properties of Samples in the System $BaTiO_3$-$Ln_2O_3.3TiO_2$, Neorg. Mat. 17 (4), (1981), 683-686
5. A. M. Gens, M. B. Varfolomeev, V. S. Kostomarov, S. S. Korovin, Crystallochemical and Electrophysical Properties of Complex Titanate of Barium and the Lanthanides, Zh. Neorg. Khim. 26 (4), (1981), 896-899
6. D. Kolar, Z. Stadler, S. Gabrščček, D. Suvorov, Ceramic and Dielectric Properties of Selected Compositions in the BaO-Nd_2O_3-TiO_2 System, Ber. Deutsch. Keram. Ges. 55 (7), (1978), 346-48
7. D. Kolar, Z. Stadler, S. Gabrščček, D. Suvorov, High Stability, Low loss Dielectrics in the System BaO-Nd_2O_3-TiO_2-Bi_2O_3, Ferroelectrics, 27 (1980) 269-272
8. K. Wakino. K. Minai, H. Tamura, Microwave Characteristics of (Zr,Sn)TiO_4 and BaO-PbO-Nd_2O_3-TiO_2 Dielectric Resonators, J. Am. Ceram. Soc. 67 (4), (1984) 278-81
9. S. Nishigaki, H. Kato, S. Yano, R. Kamimura, Microwave Dielectric Properties of (Ba,Sr)O-Sm_2O_3-TiO_2 Ceramics, Am. Ceram. Soc. Bull. 66, (1987) 1405-1410
10. L. P. Mudrolubova, K. E. Lisker, B. A. Rotenberg, T. F. Limarii, A. N. Borsch, Ceramic Materials Based on $BaLn_2Ti_4O_{12}$ Compounds for High Permittivity Capacitors, El. Teh., Ser. Radio detajli i Radiokomp. 46 (1), (1982) 3-8
11. Z. M. Durand, J. P. Boilot, Microwave Characteristics of BaO-Bi_2O_3-TiO_2-Nd_2O_3 Dielectric Resonators, J. Mat. Sci. Letters, 6 (1987), 134-136
12. W. Wersing, in Electronic Ceramics (ed. B. C. H. Steele) Elsevier Sci. Pub., 1991, 79-83
13. T. Itoh, R. S. Rudokas, New Method for Computing the Resonator Frequencies of Dielectric Resonators, IEEE Trans., MTT, MTT-25 (1), (1977) 52-54
14. M. Valant, D. Suvorov, D. Kolar, High Permittivity Microwave Ceramics - The Influence of $Ba_{4,5}Gd_9Ti_{18}O_{54}$ Addition on the Temperature Stability of Dielectric Properties, Proceedings of Electroceramics IV, Vol 1 (1994), 69-72

A Voltage-Controlled Capacitor for Wide-Band Tuning

Kenneth D. Cornett, E.S. Ramakrishnan, Gary Shapiro, Ray Caldwell,
Wei-Yean Howng, Motorola Inc., Ceramics Technology Research Lab,
Albuquerque, NM 87113

ABSTRACT:

A voltage-controlled capacitor using an MIS-like device structure has been demonstrated, and is capable of a larger capacitance tuning range than conventional varactor diodes. Use of a linear (non-ferroelectric) dielectric film provides good performance and temperature stability at RF and Microwave frequencies, and its higher dielectric constant allows a larger tuning range than can be obtained using comparable thicknesses of silicon oxide or nitride. Device characteristics can be tailored for different applications by suitable choice of a few design parameters.

INTRODUCTION:

Techniques for electrical tuning of basic radio circuits such as filters and oscillators allow radio designers to implement a wide variety of both convenient and sophisticated circuits for RF selectivity and frequency synthesis. These functions are now, and will continue to be, at the heart of wireless communications equipment.

The most common component in current use for providing of electrical tunability is the so-called variable reactance (or varactor) diode. The fundamental semiconductor device physics of these specialized diodes is well known[1]: a "reverse" bias voltage (diode "turned off") is used to control the size of the region around the p-n junction that becomes depleted of free-carriers, and functions as a variable-width dielectric. These specialized diodes are designed to both minimize the series resistance and optimize the capacitance-voltage (C-V) characteristic for a particular application. Generally speaking, a larger capacitance tuning range and/or a lower control voltage require doping levels that result in higher series resistance.

Another type of voltage-controlled capacitance can be realized in a metal-insulator-semiconductor device structure, of which the most common system by far is metal (or doped poly-Si) - silicon oxide - silicon. The device physics of this structure is also well-known[1], where a depletion region is induced at the semiconductor surface by the application of a suitable bias voltage on the metal electrode. In this case, the variable-width depletion is electrically in series with the insulator capacitance and can vary in width from zero to a maximum value (limited by the onset of inversion at the semiconductor surface, which is a function of the doping). The capacitance tuning ratio of this device is given by:

$$\frac{C_{\max}}{C_{\min}} = 1 + \left(\frac{\varepsilon_{ins} \cdot W_{d,\max}}{\varepsilon_d \cdot W_{ins}} \right) \qquad (1)$$

where ε_{ins} is the permittivity of the insulator film, ε_d is the permittivity of the depletion layer, $W_{d,max}$ is the maximum depletion width, and W_{ins} is the insulator thickness.

We realized that for RF applications, a suitably specialized M-I-S device structure[2] would have certain advantages over the conventional varactor diode for many RF tuning applications, including low control voltage, high capacitance tuning ratio, reasonable Q, and the maintenance of a capacitive characteristic under both forward and reverse voltage swings. This set of features is achieved using the novel combination of an insulator film composed of a linear (i.e. non-ferroelectric) dielectric material of high permittivity (ε_r greater than that of the semiconductor) and a thin semiconductor surface layer (in which the depletion layer may be formed) for minimum series resistance.

A) Varactor Diode

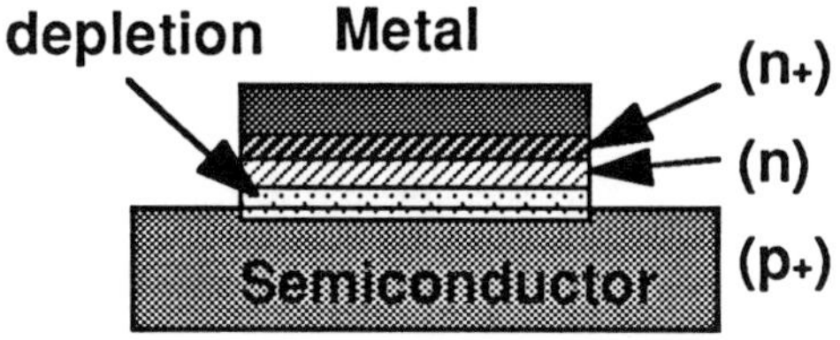

B) M-I-S Voltage Controlled Capacitor

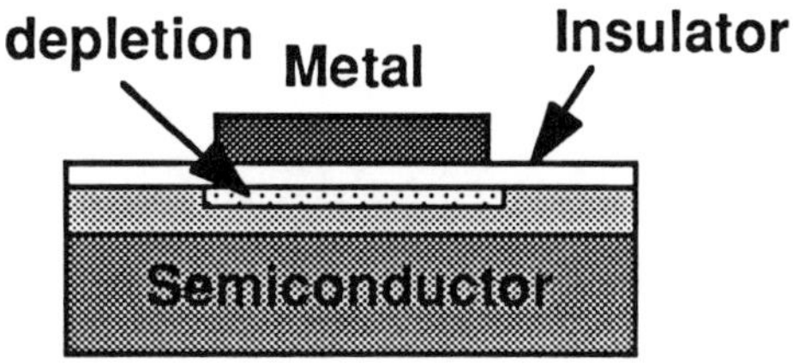

Figure 1. Comparison of (A) "typical" varactor, and (B) M-I-S voltage-controlled capacitor device structures.

RESULTS:

We have fabricated M-I-S voltage-controlled capacitors as described above using deposited Zirconium Titanate ($ZrTiO_4$) as the insulating dielectric. In bulk form, $ZrTiO_4$ is a low loss RF/microwave dielectric materials with a dielectric constant (relative permittivity) of 37-40, and a low temperature coefficient of resonant frequency. It is a simple, single-phase two component material which can be readily deposited with a very fine grain-size (<300Å) that is suitable for fabricating very fine features and therefore, high levels of device integration. $ZrTiO_4$ is compatible with standard IC fab processes and equipment, and can be deposited directly onto silicon without buffer or intermediary layers. Although Zr is not typically found in IC fabs, it behaves chemically similar to Ti in regards to its reaction with silicon. Ti is used extensively for barriers and contacts and does not contaminate active devices.

Voltage-controlled capacitors have been fabricated on both p-type and n-type surfaces, with tuning ranges appropriate to the surface dopings. As with the varactor diode, larger tuning ranges are obtained on less heavily doped wafer (higher resistivity) surfaces of either type.

Some of the radio applications for which this type of voltage-controlled capacitor may be useful include electronically tuned capacitor switches[3], integrated distributed inductive-capacitive networks[4], and integrated tunable resonators for use in oscillators and filters[5].

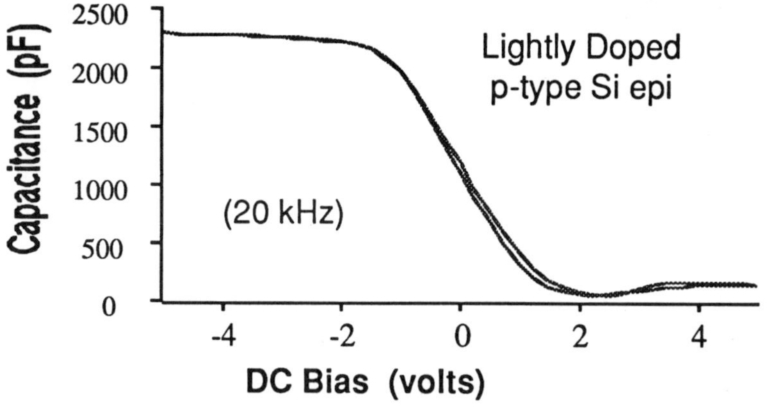

Figure 2. P-type voltage-controlled capacitor with very high tuning ratio.

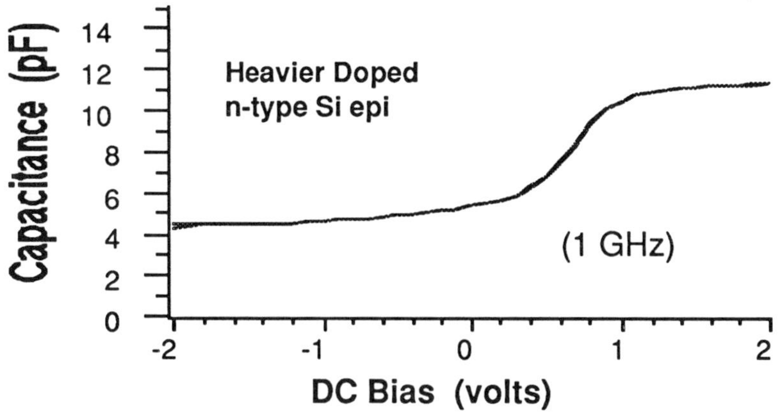

Figure 3. Smaller voltage-controlled capacitor on n-type Si for RF/microwave circuits.

REFERENCES:

1) S. M. Sze, Physics of Semiconductor Devices, Wiley & Sons, NY, NY 1981.

2) Kenneth D. Cornett, et al, "Voltage Variable Capacitor", U.S. Patent #5,173,835.

3) Branko Avanic, et al, "Electronically Tunable Capacitor Switch", U.S. Patent #5,166,857.

4) Robert E. Stengel, "Integrated Distributed Inductive-Capacitive Network", U.S. Patent #5,283,462.

5) Branko Avanic, Bob, Benenati, "Integrated Tunable Resonators For Use In Oscillators and Filters", U.S. Patent #5,166,646.

THE STRUCTURE AND MICROWAVE DIELECTRIC PROPERTIES OF CERAMICS BASED ON $Ba_{3.75}Nd_{9.5}Ti_{18}O_{54}$

F. Azough, P. Setasuwon and R. Freer
Materials Science Centre
University of Manchester/UMIST
Grosvenor Street, Manchester, M1 7HS, UK

ABSTRACT

Ceramics of $Ba_{3.75-x}Bi_xNd_{9.5}Ti_{18}O_{54+y}$ with $0\leq x\leq 0.45$ ("Bi-substituted") and $Ba_{3.75}Nd_{9.5}Ti_{18}O_{54}$ plus 1.0-4.0 wt% Bi_2O_3 ("Bi-added") were prepared by the mixed oxide route. Sintering at temperatures of 1350-1500°C yielded products with densities up to 5.6 gm cm^{-3}. TEM electron diffraction studies of Bi-free samples confirmed the space group to be Pnam; the cell parameter in the c-direction was doubled. High resolution X-ray diffraction studies enabled the structure to be refined and site occupancies identified; three primary cation sites were disordered in Ba/Nd. Microwave dielectric properties at 3GHz showed that "substitution" of Bi ($0\leq x\leq 0.35$) yielded εr values of 80, increased TCC from -80 (x=0) to -40 ppm/°C when x=0.35, but decreased Q from 2400 (x=0) to ~ 2000 (for x > 0.15).

INTRODUCTION

The growing demand for high quality ceramics suitable as resonators and other components in communication systems that operate in the 0.4 - 30 GHz band has stimulated research and development in five main families of dielectrics [1-3]. The need for high permittivity (ϵ_r), hence smaller size, dielectric of modest Q value for applications in the 0.4 to 3 GHz band renewed the interest in materials with ϵ_r > 50 [2]. Ceramics in the system $BaO-Ln_2O_3-TiO_2$ (where Ln = rare earth element) that have compositions close to $BaO.Ln_2O_3.4TiO_2$ (the so-called "1-1-4" type compounds) exhibit high relative permittivity (typically 80-110), low losses (Q values ~ 6000 to 9000 at 1 GHz) and temperature-stable dielectric properties. Early work focussed on the neodymia analogue $BaO-Nd_2O_3-TiO_2$ [4, 5], but subsequently interest developed in materials including Pr_2O_3 [6], Sm_2O_3 [7] and La_2O_3 [8]. Negas et al. [2] critically reviewed the literature on temperature compensated dielectrics, and more recently Kolar and co-workers [9, 10] reported data for a number of 1-1-4 isotypical compounds. The powder processing route and additives have a dramatic effect on the dielectric properties at microwave frequencies [2, 10].

In spite of the potential importance of 1-1-4 type and related compounds, there has been considerable ambiguity concerning the detailed structure and space group. Refinement of the crystal structure of $Ba_{3.75}Pr_{9.5}Ti_{18}O_{54}$ by X-ray diffraction suggested the space group to be Pba2 or Pbam [6]. Investigations of specimens nominally $BaO.Nd_2O_3.5TiO_2$ and $BaO.Nd_2O_3.4TiO_2$ suggested space groups of Pbam and Pba2 respectively [11, 9]. However, a recent electron diffraction study of the Pr 1-1-4 ceramic $Ba_{4.5}Pr_9Ti_{18}O_{54}$ by Azough et al. [12] revealed the space group to be Pnam.

The present study is concerned with ceramics based on $15BaO.19Nd_2O_372TiO_2$. Following confirmation of the space group, the structure of $Ba_{3.75}Nd_{9.5}Ti_{18}O_{54}$ was refined by high resolution X-ray diffraction. The effect of additions of Bi_2O_3 (in the form of the single oxide, rather than $Bi_2O_3.2TiO_2$ used by Valant et al [10]) on the microwave dielectric properties of $Ba_{3.75}Nd_{9.5}Ti_{18}O_{54}$ has also been investigated.

EXPERIMENTAL

Starting materials were high purity (> 99.5%) powders of $BaCO_3$, Nd_2O_3 and Bi_2O_3 (from Fluka Chemicals, UK) and TiO_2 (from Aldrich Chemicals UK). Powders were combined in appropriate ratios to yield two series of samples based on the composition $Ba_{3.75}Nd_{9.5}Ti_{18}O_{54}$:

(i)	$Ba_{3.75-x}Bi_xNd_{9.5}Ti_{18}O_{54+y}$	where x = 0, 0.15 0.25 and 0.35
(ii)	$Ba_{3.75}Nd_{9.5}Ti_{18}O_{54}$	plus 1.0, 2.0 and 4.0 wt% Bi_2O_3 (approximately equal to x = 0.15, 0.32, and 0.64).

Thus in the first series Bi was "substituted" for Ba, whilst in the second Bi_2O_3 was "added" to the base composition. Each batch of powder was ball milled (wet, with zirconia media; typically 40 gm powder with equal amount of milling media plus 40 cc propan-2-ol) for 8 hours, calcined at 1275°C for 4 hours, and milled again for 8 hours. The resulting slurries were dried and pressed (at 120 MPa) into pellets of 10 mm diameter and length in the range 3-6mm. Pellets were sintered in air at temperatures in the range 1350°C-1500°C for 4 hours and cooled at 120°C/hour.

X-ray diffraction analysis, using CuK_α radiation and a Philips diffractometer with horizontal goniometer was employed to investigate calcined powders and sintered specimens. Densities were determined by an immersion method. Product morphologies were examined by optical microscopy and scanning electron microscopy (Philips 505 and 525 instruments) on polished and etched surfaces.

Selected specimens were examined by transmission electron microscopy (TEM). Discs of ceramic were thinned mechanically and then ion beam thinned (ION TEC 791) to yield specimens of 3mm diameter. A Philips EM430 TEM was used for lattice image and electron diffraction studies.

Samples of the base composition were investigated by high resolution X-ray diffraction (Synchrotron Radiation Source, EPSRC Daresbury Laboratory, Warrington, UK). Spectra were collected using radiation of 1.3997 Å and 1.999241 Å for periods of 10 hours. Lattice parameters, atomic positions and occupancies of the Ba/Nd sites were determined.

Dielectric properties were determined at 3 GHz using a network analyser and sweep oscillator (Hakki and Coleman [13], post resonator method). Additional measurements of capacitance were performed at 100 kHz over the range 20-100°C to yield values of temperature coefficient of capacitance (TCC).

RESULTS AND DISCUSSION

Physical Properties

X-ray diffraction analysis revealed a primary phase with an orthorhombic structure in all specimens. When large amounts of Bi_2O_3 were included in the starting mixtures ($x > 0.35$), there was evidence of a second phase, but peaks were not sufficiently strong to enable formal indexing.

Densities were typically 5.5 to 5.6 gm cm^{-3}. The highest densities were achieved in specimens substituted with x = 0.25 - 0.45 Bi, which had been sintered at 1350°C to 1400°C; specimens with Bi "added" achieved slightly lower densities even after firing at higher temperatures (Figure 1). For simplicity, most of the data reported in this paper will be for Bi "substituted" specimens sintered at 1400°C and Bi "added" specimens sintered at 1450°C.

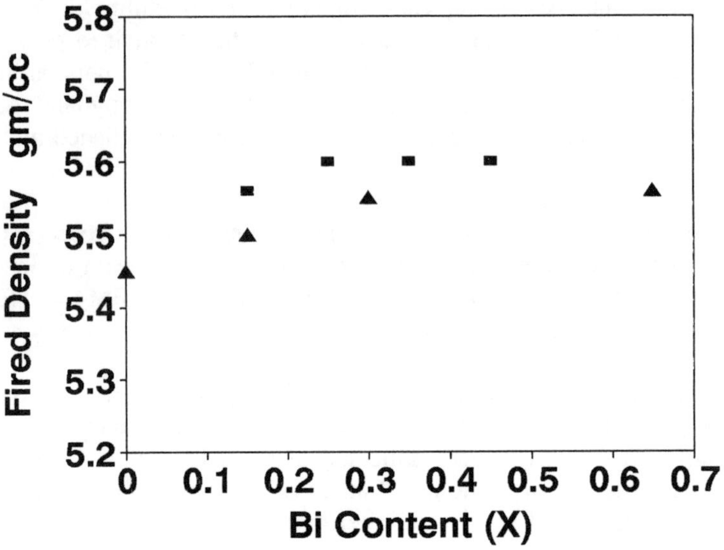

Figure 1: Density as a function of bismuth content in starting mixture for $Ba_{3.75}Nd_{9.5}Ti_{18}O_{54}$ ceramics containing substituted Bi, sintered at 1400°C (■) and added Bi, sintered at 1450°C (▲).

Structure

Optical micrographs (Figure 2) confirmed that the microstructures were very similar to those reported by other workers for BaO-Nd_2O_3-TiO_2 type materials (e.g. Valent et al. [10]) containing grains of variable size (2-10μm) and shape. For "Bi substituted" specimens prepared with mixes containing x=0.15, there is clearly evidence of porosity in the sintered products (Figure 2a); for powder batches with x=0.35 there is little evidence of porosity (Figure 2b) in the higher density ceramics. The presence of Bi_2O_3 and the elevated sintering temperatures almost certainly encouraged a liquid phase sintering mechanism. However, there was only clear evidence of melting and the formation of a Bi-Ti-Nd-rich second phase in specimens prepared with larger amounts of Bi, i.e. when x $\geq$ 0.45 (Figure 2c). It is probable that there was some vapourisation of Bi_2O_3 from all specimens during sintering. Contamination by ZrO_2 could not be detected and milling trials suggested that the weight loss from the media was $\leq$ 0.2 wt% even in extended runs.

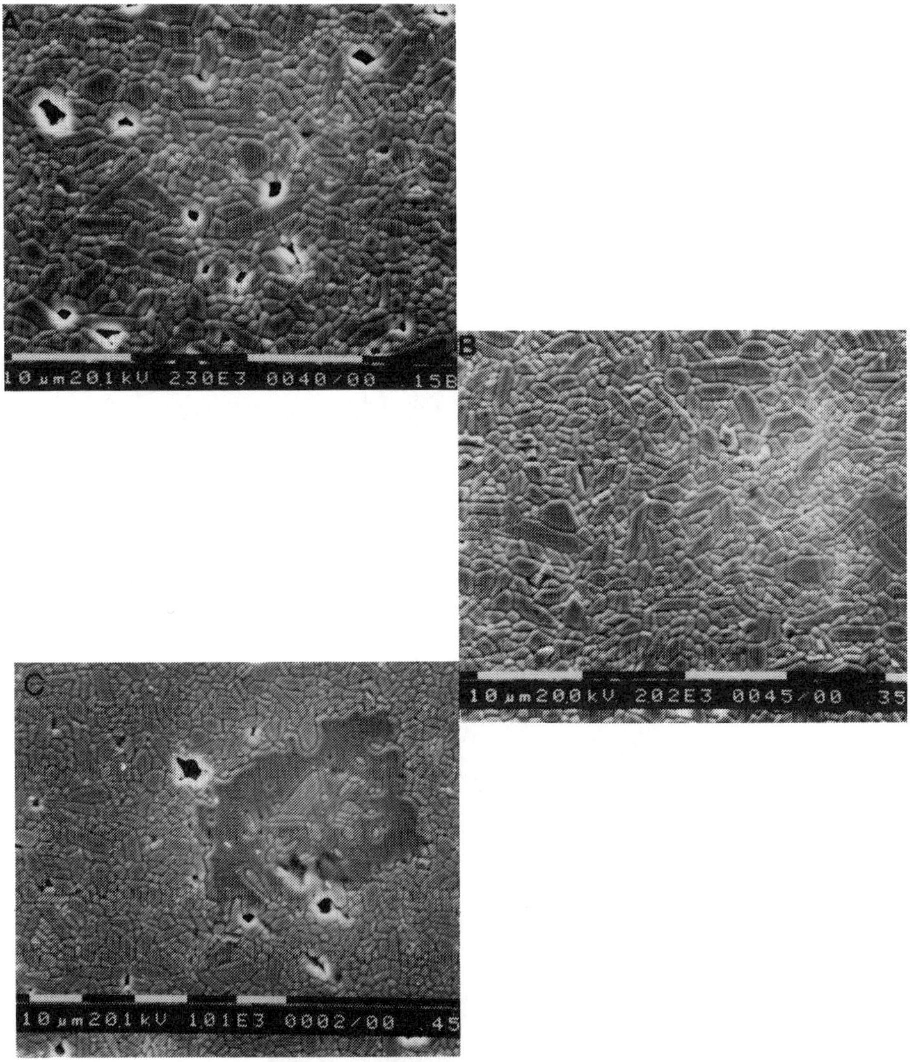

Figure 2: Scanning electron micrographs of "Bi substituted" products prepared from batches containing Bi; (a) x=0.15, (b) 0.35 and (c) x=0.45.

Figure 3 shows TEM electron diffraction patterns for the principle zone axes of undoped $Ba_{3.75}Nd_{9.5}Ti_{18}O_{54}$ ceramic. These patterns reveal that the c-axis is doubled (~ 3.8 x 2 Å) and the approximate lattice parameters for the a and b axes are 22.2Å and 12.2Å respectively. The electron diffraction patterns are almost identical to those obtained by Azough et al. [12] for $BaO.Pr_2O_3.4TiO_2$ ceramic, for which it was confirmed that the space group was Pnam. This space group is related to Pbam (one of the two space groups originally proposed by Matveeva et al. [6]; Kolar et al [9] interpreted the structure of single crystal $Ba_4Nd_{9.33}Ti_{18}O_{54}$ as orthorhombic, with space group Pba2.

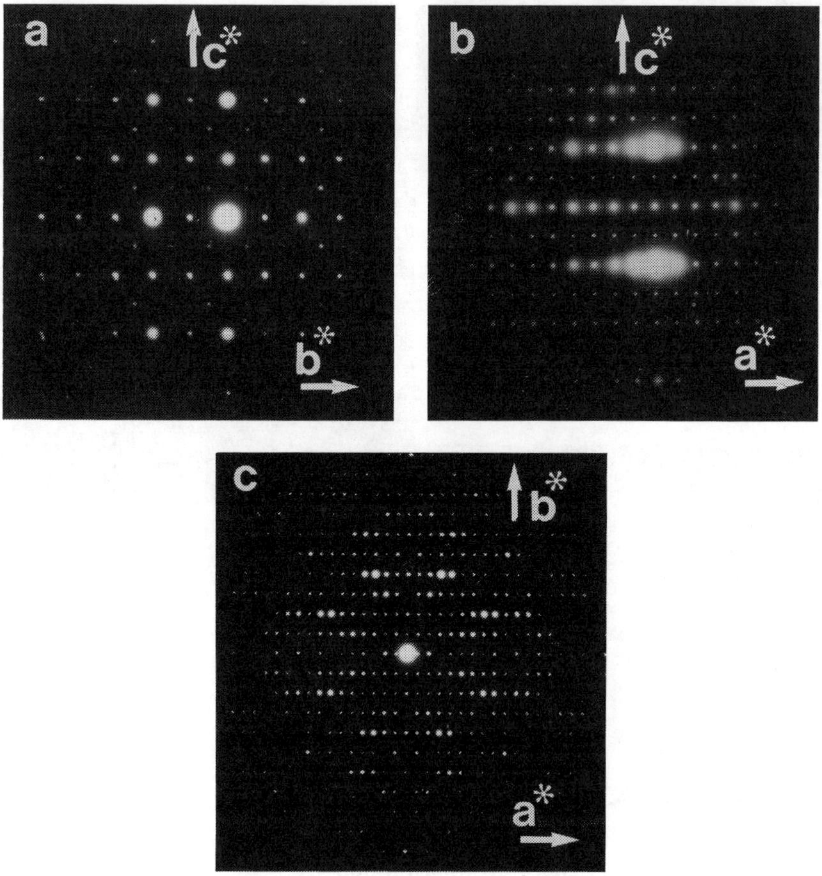

Figure 3: High resolution TEM electron diffraction patterns for the (a) [100], (b) [010] and (c) [001] zone axes of $Ba_{3.75}Nd_{9.5}Ti_{18}O_{54}$ ceramic.

Refinement of high resolution X-ray diffraction spectra for $Ba_{3.75}Nd_{9.5}Ti_{18}O_{54}$ was attempted in terms of all the available space groups. Full refinement could only be achieved using space group Pnam (space group number 62). The temperature factors were fixed for the titanium atoms; the refinement program yielded Ba:Nd ratios for site occupancies. Lattice parameters for $Ba_{3.75}Nd_{9.5}Ti_{18}O_{54}$ were deduced to be:

$$a = 22.3475\ (2)\ \text{Å}$$
$$b = 12.1938\ (1)\ \text{Å}$$
$$c = 7.6771\ (1)\ \text{Å}$$

with cell volume $v = 2092.01\ (4)\ \text{Å}^3$

Table 1 and Table 2 shows the positions of the atoms and Ba/Nd site occupancies.

Table 1: Ba/Nd site ordering

	Ba		Nd	
1	4.7(4)	:	2.8(4)	8i gen position
2	2.2(4)	:	5.8(4)	8i gen position
3	1.6(3)	:	2.0(3)	4f gen position
4	all Nd			

The coordinates of the Ba/Nd sites are broadly similar to those reported by Matveeva et al. [6] for $BaO_{3.75}Pr_{9.5}Ti_{18}O_{54}$. The site occupancies indicate that for the Ba-Nd-Ti-O compound, Nd fills site '4', dominates site '2', is the major occupant in site '3', whilst Ba is the primary occupant of site '1'. A projection of the structure of $Ba_{3.75}Nd_{9.5}Ti_{18}O_{54}$ in the (001) direction is shown in Figure 4.

The structure contains TiO_6 octahedra joined through there shared vertices to form infinite perovskite like columns along [001]. Connections of the octahedra to form a three dimensional framework are in a pattern characteristic of tungsten bronzes, with the larger cations (Ba and Nd in this case) in the cavities (Matveeva et al. [6]). The present study suggests a more complex distribution of cations than that proposed by Kolar et al. [9]. However, there is a strong suggestion that the space group varies with composition in $BaO\text{-}Ln_2O_3\text{-}TiO_2$ materials [12], and possibly with temperature, and this may go some way to explaining apparent structural differences reported in the literature.

Table 2 :	Atom coordinates and site occupancies for $Ba_{3.75}Nd_{9.5}Ti_{18}O_{54}$			
	x	y	z	B(ISO)
Ba/Nd1	0.3068 (1)	0.0910 (2)	0.2383 (8)	0.68(3)
Ba/Nd2	0.0490 (1)	0.2011 (2)	0.2429 (9)	0.68(3)
Ba/Nd3	0.5000	0.0000	0.2360(10)	0.68(3)
Ba/Nd4	0.3779 (1)	0.4070 (2)	0.2532 (9)	0.68(3)
Ti1	0.50000	0.50000	0.00000	0.25
Ti2	0.4377(9)	0.198(2)	0.00000	0.25
Ti3	0.111(1)	0.400(2)	0.00000	0.25
Ti4	0.1633(9)	0.110(2)	0.00000	0.25
Ti5	0.263(1)	0.335(2)	0.00000	0.25
Ti6	0.50000	0.50000	0.50000	0.25
Ti7	0.434(9)	0.199(1)	0.50000	0.25
Ti8	0.107(1)	0.401(2)	0.50000	0.25
Ti9	0.1647(8)	0.115(2)	0.50000	0.25
Ti10	0.260(1)	0.336(2)	0.50000	0.25
01	0.1579(9)	0.105(2)	0.251(1)	5.7(1)
02	0.195(1)	0.412(4)	0.00000	5.7(1)
03	0.236(2)	0.179(3)	0.00000	5.7(1)
04	0.191(1)	0.411(4)1	0.50000	5.7(1)
05	0.239(2)	0.178(3)	0.50000	5.7(1)
06	0.2776(8)	0.321(2)	0.251(1)	5.7(1)
07	0.020(1)	0.377(5)	0.00000	5.7(1)
08	0.020(2)	0.359(6)	0.50000	5.7(1)
09	0.4430(9)	0.216(2)	0.250(1)	5.7(1)
010	0.117(3)	0.243(2)	0.00000	5.7(1)
011	0.115(3)	0.244(3)	0.50000	5.7(1)
012	0.50000	0.50000	0.250(1)	5.7(1)
013	0.0839(6)	0.015(3)	0.00000	5.7(1)
014	0.361(2)	0.249(5)	0.00000	5.7(1)
015	0.484(2)	0.344(1)	0.00000	5.7(1)
016	0.421(2)	0.048(2)	0.00000	5.7(1)
017	0.322(2)	0.451(3)	0.00000	5.7(1)
018	0.075(1)	0.064(4)	0.50000	5.7(1)
019	0.363(2)	0.273(5)	0.50000	5.7(1)
020	0.470(3)	0.352(2)	0.50000	5.7(1)
021	0.403(2)	0.059(2)	0.50000	5.7(1)
022	0.306(2)	0.467(3)	0.50000	5.7(1)
023	0.0985(9)	0.398(2)	0.250(1)	5.7(1)

R factors

R(I) - 8.82%; R(WP) - 15.98%; R(Expected) - 13.83%

Figure 4: Projection of the structure of $Ba_{3.75}Nd_{9.5}Ti_{18}O_{54}$ in the (001) direction (↑y →x). Larger cations in the cavities represent disordered Ba/Nd sites; smaller cations in the cavities represent ordered Nd sites.

Dielectric Properties

Figure 5 shows the relative permittivity of $Ba_{3.75}Nd_{9.5}Ti_{18}O_{54}$ ceramics at 3 GHz as a function of Bi content in the starting batch. From the ϵ_r value of 78 in the undoped specimens, the relative permittivity increased slightly to 80 in the Bi-substituted specimens. In contrast, the relative permittivity decreased to 76 $\pm$ 1 in the Bi-added specimens, reflecting the lower densities (Figure 1).

In the undoped Barium neodymium titanate the dielectric Q value at 3 GHz was ~2400. Specimens prepared from "substituted" Bi batches with x $\leq$ 0.15 had similar Q values, but higher substituted levels led to a reduction in Q value to approximately 2000 (Figure 6). In contrast specimens prepared from "added" Bi batches at 0.15 to 0.35 atomic fraction exhibited a slight increase in the Q value, but again greater concentrations of Bi caused Q to fall. The relative permittivities and Q values for specimens prepared with little or no Bi are comparable with the results reported by Valant et al [10] after the reagent grade starting powders had been calcined, leached, precipitated and small amounts of Mn and WO_3 had been added.

Furthermore, Valant et al. [10] noted the beneficial effect of Bi additions (in terms of $Bi_2O_3.2TiO_2$), such that at 6.5 wt% Bi_2O_3, Q = 5800 (at 1 GHz), ϵ_r ~88 and the temperature coefficient of capacitance (TCC) was close to zero. The earlier study of Kolar et al. [14] with $Ba_{2.25}Nd_{4.5}Ti_9O_{27}$ + 10 wt% $Bi_2O_3.2TiO_2$ obtained a TCC value of -24 ppm/°C at 1 MHz. Various studies on $BaNd_2Ti_4O_{12}$ (the 1-1-4 composition) at 1 MHz suggested TCC values of -113 to -160 ppm/°C [15]. The temperature dependency of permittivity (TCC) for the samples investigated in the present study are shown in Figure 7. In the undoped specimens TCC was -80 ppm/°C, but decreased steadily to -40 ppm/°C in the samples "substituted" with 0.35 atomic fraction of Bi. "Addition" of Bi also improved the TCC to -40 ppm/°C, which was independent of the amount of bismuth. It is tempting to suggest that "substitution" of more than 0.35 atomic fraction Bi should yield ceramics with TCC values close to zero. Unfortunately, such specimens are likely to have Q values less than 2000 at 3 GHz (Figure 6). However, small amounts of Bi substituted into $Ba_{3.75}Nd_{9.5}Ti_{18}O_{54}$ are clearly helpful and the optimum level is defined by the trade-off in TCC and Q value.

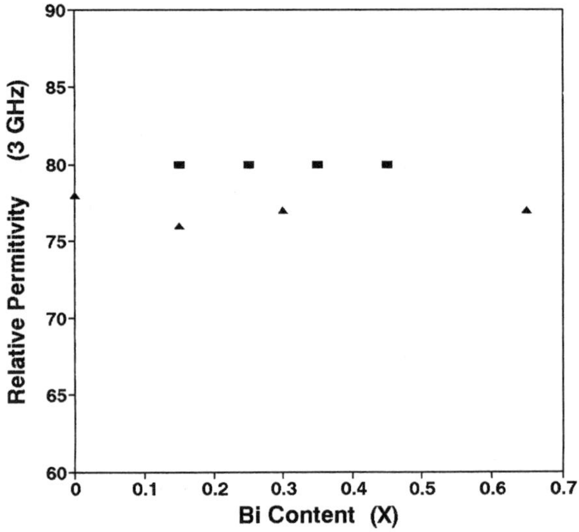

Figure 5: Relative permittivity (at 3 GHz) for $Ba_{3.75}Nd_{9.5}Ti_{18}O_{54}$ ceramics as a function of Bi content in starting mixes (Bi "substituted" samples ■; Bi "added" samples ▲).

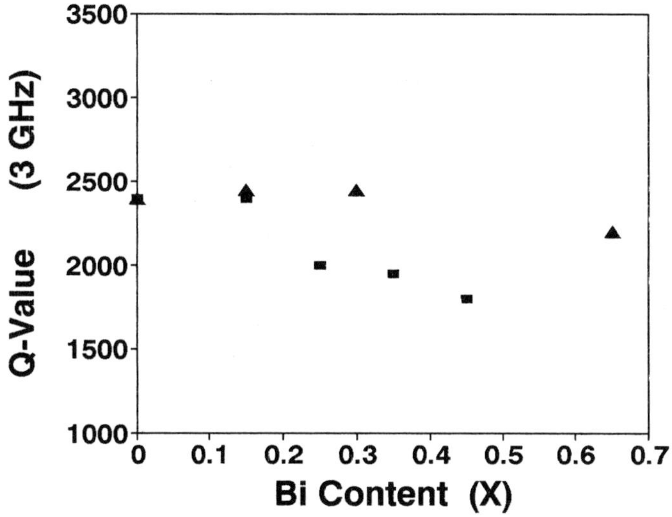

Figure 6: Dielectric Q values of (at 3 GHz) for $Ba_{3.75}Nd_{9.5}Ti_{18}O_{54}$ ceramics as a function of Bi content (Bi "substituted" samples ■; Bi "added samples ▲).

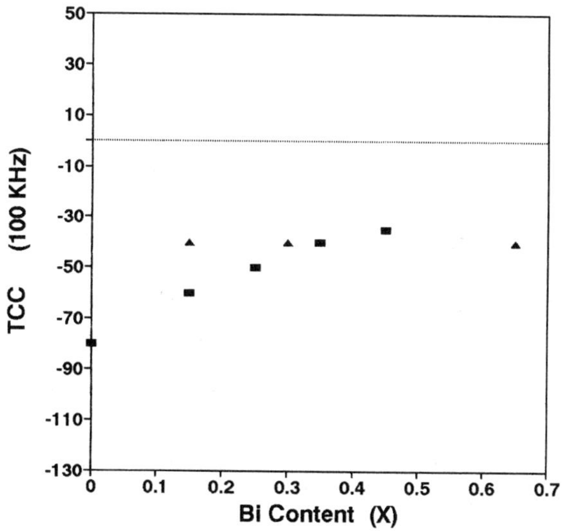

Figure 7: Temperature coefficient of capacitance (ppm/°C) at 100 kHz of $Ba_{3.75}Nd_{9.5}Ti_{18}O_{54}$ ceramics as a function of Bi content in the starting batch (Bi substituted samples ■; Bi "added" samples ▲).

CONCLUSIONS

Good quality ceramic based on $Ba_{3.75}Nd_{9.5}Ti_{18}O_{54}$ was prepared by the mixed oxide route. The inclusion of up to 0.35 atomic fraction Bi in the starting batches improved densification. High resolution X-ray diffraction studies of Bi-free specimens enabled the structure to be refined on the basis of space group Pnam; the cell parameter in the c-direction was shown to be doubled. Three of the primary cation sites were disordered in Ba/Nd, whilst the fourth contained only Nd.

"Substitution" of Bi for $x \leq 0.35$ yielded specimens with relative permittivities of 80 at 3 GHz; Bi-"added" specimens had lower ϵ_r values of 76 ± 1, due to the lower densities. Q values of 2400 at 3GHz were achieved in ceramics "substituted" with Bi for $x \leq 0.15$, and Bi "additions" with $x \leq 0.35$. TCC increased from -80 to -40 ppm/°C as Bi substitution increased from $x = 0$ to 0.35; in Bi-added samples TCC = -40 ppm/°C for $0.15 \leq x \leq 0.6$.

ACKNOWLEDGEMENTS

We gratefully acknowledge the help of Dr. A. Bell of EPSRC Daresbury Laboratory with the structural refinement.

REFERENCES

1. K. Wakino, "High frequency dielectrics and their applications", Proceedings of 6th IEEE Internat. Symp. on Applications of Ferroelectrics, Bethlehem Pa., June, 1986, pp.97-106 (1986).

2. T. Negas, G. Yeager, S. Bell and R. Amen, "Chemistry and properties of temperature compensated microwave dielectrics", NIST special publication 804, Chemistry of Electronic Materials (Proceedings of Intl. Conf. held Jackson, WY, Aug. 1990) pp.21-33 (1991).

3. R. Freer, "Microwave Dielectric Ceramics - an Overview", Silicate Industriels, 9-10, 191-197 (1993).

4. R.L. Bolton, Dissertation, University of Illinois (1968).

5. D. Kolar, Z. Stadler, S. Gabrščček and D. Suvorov, "Ceramic and dielectric properties of selected compositions in the $BaO-TiO_2-Nd_2O_3$

system", Ber. Deut. Keram. Ges., 55, 346-348 (1978).

6. R.G. Matveeva, M.B. Varfolomeev and L.S. Il'yusheenko, "Refinement of the composition and crystal structure of $Ba_{3.75}Pr_{9.5}Ti_8O_{54}$", Trans. from Zh. Neorg. Khimi., 29, 31-34 (1984).

7. S. Nishigaki, H. Kato, S. Yanoan, R. Kaimimura, "Microwave dielectric properties of $(Ba,Sr)O\text{-}Sm_2O_3\text{-}TiO_2$ ceramics", Am. Ceram. Soc. Bull., 66, 1405-1411 (1987).

8. J.P. Guha, "Synthesis and characterisation of barium lanthanum titanates", J. Am. Ceram. Soc., 74, 874-880 (1991).

9. D. Kolar, S. Gabrŝĉek and D. Surorov, "Structural and dielectric properties of perovskite-like rare-earth titanates", Proceedings Third Euro-Ceramics Conference, eds. P. Duran and, J.F. Fernandez (Faenza Editrice Iberica S.L. 1993) pp.229-234.

10. M. Valant, D. Suvorov and D. Kolar, "Microwave ceramics of high dielectric constant", ibid, pp.235-240.

11. R.S. Roth, F. Beach, A. Santoro and K. Davis, "Structure of the non-stoichiometric solid solutions $Ba_2RE_4(Ba_x+RE_{2/3-2/3x})Ti_9O_{27}$ (RE=Nd,Sm)", Abstract 07.9-9, Fourteenth International Congress of Crystallography, Perth Australia, 12-20 August 1987.

12. F. Azough, P.E. Champness and R. Freer, "Determination of the space group of ceramic $BaO.Pr_2O_3.4TiO_2$ by electron diffraction", submitted to J. Applied Crystallography.

13. B.W. Hakki and R.D. Coleman, "A dielectric resonator method of measuring inductive capacitance in the millimeter range", IRE Trans. on MTT, MTT-8, 402-410 (1960).

14. D. Kolar, S. Gabrŝĉek, Z. Stadler and D. Suvorov, "High stability low loss dielectrics in the system $BaO\text{-}Nd_2O_3\text{-}TiO_2\text{-}Bi_2O_3$", Ferroelectrics, 27, 269-272 (1980).

15. D. Kolar, "Chemistry and properties of perovskite-type rare-earth titanates for microwave applications", Proceedings of ELECTROCERAMICS IV, (Aachen, Sept. 1994) eds. R. Waser et al (Verlag der Augustinus Buichhandlung, 1994) pp.3-10.

Keyword and Author Index